Multimodal Transport Systems

Multimodal Transport Systems

Edited by
Slim Hammadi
Mekki Ksouri

iSTE

WILEY

First published 2014 in Great Britain and the United States by ISTE Ltd and John Wiley & Sons, Inc.

ISTE Ltd
27-37 St George's Road
London SW19 4EU
UK

www.iste.co.uk

John Wiley & Sons, Inc.
111 River Street
Hoboken, NJ 07030
USA

www.wiley.com

Library of Congress Control Number: 2013947315

British Library Cataloguing-in-Publication Data
A CIP record for this book is available from the British Library
ISBN: 978-1-84821-411-8

Table of Contents

Preface

Transport engineering is the set of functions characterizing the specifications of user needs, the study and design of technical and industrial solutions, as well as the implementation and monitoring of these solutions within the field of transport.

A user of a transport system could be a network controller or a traveller who requires information for his journey. In all cases, the idea is to provide a "client" with advanced mobility services.

Transport engineering is a field of interest for researchers all over the world. Their work aims to improve existing solutions or bring new and effective solutions to functional problems.

New methods for solutions have been developed relative to the emerging needs of users as well as to recent communication technologies and travel within an ergonomic, socio-economic and ecological context.

It is due to the grouping of the best research teams within CISIT, the International Campus for Security and Intermodality of Transport[1], and the innovation of the northern French region of Nord Pas de Calais that numerous long-term partnerships have been developed between the industry and transport operators, as well as international researchers who have been invited to join. From these partnerships,

---

1 http://www.cisit.org.

rich in research and innovation, the idea of writing a reference book was born in order to give value to the results obtained.

These works refer to the CISIT project “Urban and inter-urbanmultimodal networks”, which was set up in order to respond to the strategic objective “Optimal management of multimodal chains”. This objective, which is associated with CISIT’s three big strategic objectives: “New challenges for clean, light and safe vehicles”, “Intelligent, safe and secure mobility” and “Morpho-adaptive security and human factors”, contributes to the challenge of sustainable mobility of people and goods.

The authors of this reference book contribute to this challenge directly: they offer the reader an update of the state of the art of all the techniques, approaches and methods for the specification, design, optimization and implementation of advanced mobility services.

We would hereby like to gratefully acknowledge all contributors to this book, for the most part members of CISIT and researchers in the various fields related to Transport Engineering and Advanced Mobility Services.

Slim HAMMADI
Mekki KSOURI
October 2013

Chapter 1

Dynamic Car-pooling

1.1. Introduction

In order to mitigate the negative impact of private cars and thus heal the environmental image of a personal vehicle, car-sharing systems were born. Within this context, car-pooling in particular has been a notable success thanks to the contributions it brings mainly by reducing the number of cars on the road. Indeed, making the personal car a common mode of transport, car-pooling plays a role in the reduction of harmful gas emission rates. The contributions are quantifiable in terms of "non-emitted" CO_2, in addition to the many advantages it offers on both an individual and a collective level (e.g. reduction of budgets allocated for transport, time-space flexibility, comfort, social balance, etc.). Thus it has made its entrance into the field of research, and numerous systems have since emerged. Several studies have been conducted in a manner that draws on the fields of computer science, artificial intelligence, GIS (Geographical Information Systems), the Internet and telecommunications, etc. Making use of new technologies and the establishment of a more or less evolved system has precedence over any other goal in existing approaches. Web-based media are today operational and allow the

---

Chapter written by Slim HAMMADI and Nawel ZANGAR.

general public to register and benefit from fairly limited services such as publications and consultation of offers and demand, as well as acquisition of contact details for potential car-poolers. Unfortunately, this type of system is the only one that deserves recognition because the rest, despite their openness to advanced features such as the integration of real time and the automation of allocation tasks based on multi-agent systems, have remained at the "*idea*" or "*draft*" stage, and are not liable to improvement. Irrevocably launched onto the road towards improvement, in this chapter we propose the implementation of an optimized dynamic car-pooling system. Two main concepts will be discussed in particular, namely the modeling of the problem as a distributed dynamic graph, on the basis of which distributed software architecture is established, and the deployment of a multitude of autonomous entities under this architecture. The combination of multi-agent systems with the foundations for optimization has thus been put in the service of effectiveness of processing for the establishment of an approach in the context of distributed artificial intelligence.

1.2. State of the art

In recent years, the sharing of vehicles has become a remarkable phenomenon mainly due to the economic and environmental benefits it offers. Thus, people tend to go for the "self-shared" vehicles (self-service vehicles, car-pooling, etc.) and discard their own automobiles. In this context, there are nowadays several studies that revolve around the sharing of vehicles, allowing the completion of relatively efficient systems to exist in about 1,000 cities around the world already [WCC]. These systems provide more or less satisfactory services for users but are still in the early stages. The work done to date can be subdivided into two categories according to the criteria for booking management adopted within the system. The first category is based on a static booking management while the second deals in real time with a dynamic aspect. In what follows, we present a non-exhaustive list of what has been done in this regard by making a distinction between:

– operational systems implementing the concept of "Shared Vehicles" (we actually focus on car-sharing, car-pooling sites remaining, for the most part, open and non-optimized systems);

– the academic work done for modeling and optimization of such systems.

The existing operational systems are:

Static systems

– SEFAGE (SElbstFAhrerGEnossenshaft), which could be translated as “drivers club”. This is the oldest car-sharing organization traceable in the literature. It was founded in Zurich, Switzerland in 1948. It was essentially a club where members came together to buy a car. Without any commercial purpose, the main objective was to offer the service of having a vehicle available when needed. As the initiators were not aware of its innovative characteristics, SEFAGE never developed further.

– Lilas [LIL]: car-sharing was strongly implemented in Lille in France through this service in which a booking is made in advance by phone or online using a member number. The member, who receives a monthly invoice, must book each time for at least one hour of service but has the freedom of choosing a car of their choice from a station of their choice. However, despite the variety of stations and vehicles it provides to customers, the Lilas system has the disadvantage of limiting the time of use of its vehicles, working in a loop (return by the user to the station of departure) and not having real-time reservation services. Lilac offers competitive pricing for users subscribed to local transport and for families, and combines well with other forms of transport.

– Modulauto is the name associated with the car-sharing service available to the inhabitants of the city of Montpellier and the town of Nîmes in France. As part of the *France Auto-partage* service, Modulauto offers its members a fleet of self-service vehicles. Users reserve a car online or by phone at least 30 minutes in advance and the vehicle must be returned to a Modulauto station.

– Mobility is the market leader for car-sharing in Switzerland, with a portfolio of 55,500 clients for which it provides 1,700 vehicles. A partnership with public transport has contributed largely to the development of Mobility.

– Communauto [BEN] in Canada (Quebec, Montreal, Sherbrooke, Gatineau): since its foundation in 1994, Communauto appears to be a pioneering enterprise in America, as the manager of the oldest and one of the most important car-sharing services that has emerged on its side of the Atlantic. Communauto, which now has more than 14,000 members in Quebec, is the first car-sharing organization in the world to have signed the Charter for Sustainable Development of the International Union of Public Transport (IUPT).

– City Car Club presents itself as the solution to problems faced by owners of private vehicles in the UK. It is a car-sharing service offered to British citizens and, like its predecessors, has a fleet of self-service vehicles that members share amongst themselves.

– Cambio Stadt [MOB] in the city of Bremen in Germany: implemented in 2002 and owing to a partnership with Vivaldi and cooperation with Civitas, during its first three years, this project experienced a 43% increase in the number of users (2,455 to 3,512 in January 2005).

– I-GO and zipcar USA: having the same principle as the majority of the various systems listed above, car-sharing in the United States presents itself, for the zipcar company founded in June 2000 [ZIP] and I-GO [IGO], as a system of advance booking of self-service vehicles of limited use duration. These vehicles are collected from stations designed for this purpose and delivered back to the starting station after use.

The static systems described above require advance booking based on static reasoning, without considering instantaneous events. Thus, these systems do not provide real-time vehicle allocation, nor do they provide an immediate response to the user. To remedy this deficit, dynamic car-sharing systems have emerged with real-time service management.

Dynamic systems

– PRAXITELE: the PRAXITELE Research and Development program was founded in 1993 based on a consortium from the industry (CGEA, Renault, the Dassault Electronics Group and EDF) and two research institutes (INRIA and INRETS). The service became operational in October 1997 in Saint-Quentin-en-Yvelines with 50 cars (PRAXICARS) and five carparks (PRAXIPARKS). PRAXITELE is a new form of public transport complementary to public transport and taxis. This new service is primarily intended for journeys and schedules when demand is diffuse and public transport is inadequate in terms of frequency and profitability. Its use is restricted to a geographical area ranging from a few hundred meters to several kilometers in some cases. To use PRAXITELE, one must first become a member of the service. A "credit card"-type memory card using contactless technology, the PRAXICARD, is then delivered to the client allowing them to access the cars, consult the information terminals installed in the stations and carry out payment. The PRAXIPARK stations act both as an interface between the system and its users through information terminals (information exchange with PRAXICENTRE, controls), and as a local operation system through its chargers (for recharging vehicles, etc.). The PRAXICAR vehicles use electric engines and are equipped with automatic chargers which charge by induction, facilitating their use, management and operation. They include an on-board computer which controls the opening in order to collect payment and ensure a dialogue with the user. Currently, the vehicles are electric versions of existing models (Renault Clio). Eventually, the idea is to use the concept of small electric cars, with dimensions better suited to urban use.

– LISELEC [LIS]: created by PSA, VIA GTI and Alcatel CGA Transport, LISELEC is part of the global travel policy that has been leading the the La Rochelle agglomeration for many years. The experiment began in September 1999 in partnership with: the General Council of Charente-Maritime, the Regional Council of Poitou-Charentes, ADEME Poitou-Charentes, EDF and PREDIT. The service offered by LISELEC works similarly to the PRAXICENTRE system mentioned above, also with 50 electric cars (106, Saxo, Berlingo and Gem) dispersed across seven stations. These vehicles are available at any time of the day or night with a membership contract. With LISELEC, parking is free in La Rochelle and in the spaces reserved in stations. This system is intended for all city residents and for all regular visitors for short urban journeys and additionally offers preferential tariffs for long-distance travel compared to a vehicle rented from a classic car hire company (National/Citer). At the end of April 2002, there were 485 members with an average increase of six new members per week.

– Cité Vu from VU LOG: VU LOG is a company offering a different take on car-sharing in downtown Antibes. It presents a new car-sharing system (Summer 2008) called "Cité Vu" which offers its members a fleet of public individual electric vehicles. These are organized into self-service with instant access and do not require prior reservation. Members of this service therefore benefit from an ease of access as well as the possibility of parking anywhere in the city without being obliged to return the car to a preset station. However, the vehicles available are only designed for short-distance urban travel, traveling at a maximum speed of 45 km/h.

– AutoLib [WCC]: in the same context as "Cité VU" by VU LOG, a new project named AutoLib was launched in December 2011. This project takes into consideration the management of reservations and customer requests in real time. AutoLib should make available to prospective members a fleet of electric vehicles, thereby reducing CO_2 and greenhouse gas emissions.

– Telebus (On-Demand Transport for the disabled [BOR 97]): this TAD system, established in Berlin, Germany in 1981, has the interesting characteristic of having been the testing ground for methods of "smart" control whose application in the years 1990/2000 has increased covered demand by 40%.

Academic points of view

The *universities of Quebec* can claim a certain leadership regarding the design of computer systems to support the decision for mobility and transport: CRT MONTREAL, GERAD, CIRRELT Laboratory, GIRO Society. They have extensively studied conventional routing problems, load planning and allocation in static and dynamic contexts [BAR 07; COR 02]: *Pick up and Delivery*, *Dial and Ride* [COR 03; COR 07; CED 01]: *Crew scheduling*, using linear programming techniques [BAR 98] (column generation, etc.) now commonly used in Air France, SNCF, EDF and France Telecom R/D.

However, these problems, since the history of Decision Support, have also been studied in Europe, particularly in Italy (La Sapienza University and I.P. Rome for the *Dynamic TAD*, I.P. MILAN for *Intelligent Car-sharing*), Belgium (ULB for *Network Design* and *Pricing*), Germany (Bonn, Berlin, Osnabruck, Saarbrücken, Kaiserslautern, etc. for the coupling between Mobility and Industrial Planning), and France (UTT, UTC HEUDYASIC: *Tours Planning*, *Network Design*; IRCYN NANTES: *Hospital Mobility*; LAGIS Ecole Centrale of Lille: *Multimodality*, and at the industrial level, in DER AIR France or SSII EURODECISION, ILOG/IBM, AERTLIS, etc.). If one of the current trends is to encourage innovative mobility and new vehicle categories (INRIA, LASMEA, CRYSTAL ANR Project), another is to consider it through a systematic perspective by linking it to the concept of sustainable economy and environmental concerns [COR 03]. Also, a request could be made to schedule days such that production operations and transport activities related to internal enterprise logistics are coupled [MCB 83; SCHU 78; POC 06].

The adaptation of algorithms from a static context (point of view of most programmers) to a dynamic context has often been studied, but most come with the cost of not considering how users, vehicles and controllers actually collaborate through the communication architecture [JES 08; LEC 06; PAS 95]. Studies on the statistical analysis of the demand for mobility were carried out on a macro level during the 1980s (Logit TERESE of CERTU model and INRETS, [TER]).

These models do not address the question of the susceptibility of demand towards a micro-system (TAD, etc.) in case of changes in its configuration [JÄG 03]. Interesting studies have been conducted in this regard at the University of Montreal, around the management of a fleet of ambulances for emergencies [BRO 03], which involves a spatio-temporal model, and at MIT on the elasticity of demand for public transport [ASH 98]. Regarding *Pricing*, the most innovative experiments have involved *Cooperative* or *Non-Cooperative Game Theory* models, intended to enable the identification of stable market shares in multimodal transport [ALT 04; BOU 02; TAM 91], as well as *bi-level optimization* models for combining pricing with a control procedure (decongestion) for access to a network.

The Optimization of Networks has been promoted widely by the deregulation of energy production markets, transport and telecommunications, initially with a focus on issues related to traffic stability and the consideration of security constraints [HU 05; WAR 52]. The emergence of embedded technologies allowing the planning of real-time networks (*SmartGrids*, etc.) is leading to a reformulation of classical problems for very large amounts of data on dynamic networks (with time indexing nodes) and stochastic demands. The case of data acquisition and monitoring systems (Sensor Networks) is an especially rich case [FLE 09].

Very recently, prospective studies were conducted in California and The Netherlands on the potential opportunities offered by the introduction of Web technology and distributed in the management of mobility support systems [VAN 04; GIL 02; HEU 08; HIG 00; STO 00], with a focus on the design of advanced system/user communication systems, the management of human resources and the control of safety. Studies have also been conducted on the prospect of the increased flow of automated vehicles in mobility and transport services [KON 06].

1.3. Complexity of the optimized dynamic car-pooling problem: comparison and similarities with other existing systems

In order to properly identify the complexity of the car-pooling problem in terms of optimization difficulty, two types of optimization problem are considered: the first concerns the combinatorial optimization problem for which there is no accurate and fast algorithm. We can cite a known problem, which is that of the traveling salesman, classified as "NP-Hard" whose exact solution can only be done by a time calculation proportional to N^n, where n is an integer and N is the number of unknowns. The second type relates to optimization problems of which the variables are continuous. For this type of problem, there are no algorithms guaranteeing certainty of finding a global optimum with a finite number of calculations.

A comparative study on the problem of car-pooling with dynamic optimization, CDO, with other optimization problems known for their complexities has been conducted. Indeed there are many similarities between the CDO dynamic car-pooling problem and the traveling salesman problem, TSP, such as the nodes (origin/destination vs. cities), and the objective function (distance vs. time, distance, CO_2 emission). The CDO is hence considered an NP-Hard combinatorial problem whose complexity is expressed in a manner similar to that of the TSP.

$$(O(CDO) \equiv O(TPSP))$$

In addition to the traveling salesman problem, we can also refer to the vehicle routing problem (VRP) and in particular the collection and delivery of objects (i.e. products, goods, people, etc.). A comparative study of the CDO problem with two variants of DPDPs (Dynamic Pickup and Delivery Problems), which are the Swapping Problem [ANI 92; COR 07], SP, was performed. This study showed that there was even more similarity than with the TSP, bearing in mind that the VRP problems are known for their high complexity of exponential order. By analogy to these problems, the CDO is thus stated as an NP-Hard combinatorial optimization problem

$$O(CDO) \equiv O(Swapping\ Problem) \equiv (Dynamic\ DARP)\,.$$

hence the great difficulty in solving the CDO problem as well as establishing an efficient optimization system.

1.3.1. *Graphical modeling for the implementation of a distributed physical architecture*

For the graphical modeling of the car-pooling system, we investigated works in the context of the vehicle routing problem [IOR 07] and in particular the problems of collection and real-time distribution [BER 10; WU 09]. Indeed, the random and dynamic aspects of events and data [SAV 95; XIA 08] defining these problems join the randomness of the supply and demand of the real-time car-pooling problem. The transport problem of on-demand transport for disabled persons [ATA 10; COR 03], which consists of adjusting the supply according to the demand of users, is considered here, especially since it is even closer to our problem, given that we place ourselves in the context of transportation of people and not of goods.

Bearing in mind the very random aspect of supply and demand by users, the architecture of the proposed system must be able to monitor developments in real-time and will be based on the concept of a dynamic distribution graph. Thus, for n users issuing car-share requests and m vehicles offering trips, whether they are already in circulation or not at a given time, *t*, we define the dynamic car-pooling system as the triplet $T(t) = (G(t), R(t), O(t))$. In this expression, each parameter is relative to a particular component that contributes to the successful implementation of the system:

– $G(t) = (N(t), A(t))$ is a dynamic directed graph representing the served network where all offers and requests of users of the service are scattered.

– $N(t)$ is the set of nodes constructed based on geographical positions of users of the service, whether as car-sharers or car-sharees, and the details of the required or expected travel (i.e. benchmarks and routes to be taken).

For reasons of safety and security, the geolocation of users (i.e. pedestrians and cars) is done in real time thanks to a positioning module integrated within the system.

$A(t)$ is the set of arcs on the graph, which are formed as a result of the extraction of required information on what is available at a given time, *t*. They essentially represent the routes that will be served by the cars and therefore correspond to the journey remaining to be traveled by drivers during the trip or not. These routes are constructed and updated in real-time based on the time when a vehicle passes a check-point chosen by the system. This representation must be consistent with the nature of the roads as well as the possibilities for using them (e.g. one-way street, two-way street, etc.). An arc can be oriented either in one direction or in both if two or more cars have used the same section of road in opposite directions.

Hence the itineraries of cars are composed of several sections. Two or more cars can share more than one road in the same direction. Thus defined, these sections are the arcs of the graph $G(t)$ connecting the nodes between them, thus identifying their orientation according to the directions in which represented roads are used. The end nodes of these segments of journeys can be the origin or destination or both at once.

Each node can have one or more successors and hence represents an intermediate or final destination of the vehicle(s) or user(s).

We define the set of successor nodes of x written $N_x^+(t)$ as follows:

$$N_x^+(t) = \{v \in N(t) / x \neq v \text{ and } (x,v) \in A(t)\} \quad [1.1]$$

Similarly, each node can have one or more predecessors which are the original ends (i.e. origins) of arcs whose destination node is considered.

The set of predecessor nodes of x, written $N_x^-(t)$, is defined as follows:

$$N_x^-(t) = \{v \in N(t) / x \neq v \text{ and } (v,x) \in A(t)\} \quad [1.2]$$

1.3.2. *Collection of requests for car-pooling and data modeling*

Given that the random nature of our graph relies heavily on data that is constantly changing, we propose in the following to see how car-pooling requests will be structured, and data modeling.

We consider $R(t) = \bigcup_{p=1}^{n} \{R_p(t)\}$ the set of n requests issued by n users of the service $U(t) = \bigcup_{p=1}^{n} \{U_p(t)\}$.

Only users logged into the system at time *t* and who have issued car-pooling requests are considered.

1.3.2.1. *Structure of a request for car-pooling*

The structuring of car-pooling requests is done as follows:

$$R_p(t)=\left\{R_p^+,R_p^-,P_p,\left[d_p,a_p\right]\right\} \quad [1.3]$$

Each request is specific to a particular user, U_p, indicating their travel preferences:

– R_p^+ indicates the geographical coordinates of the desired origin of travel. If the user fails to specify their point of departure, geographical positioning at time t is captured and will be considered the original source of the route to follow.

– R_p^- is the final destination that the user wishes to achieve. Introduced in a similar manner as at the beginning of the planned route, the destination can be inserted as an address or geographic coordinates.

The origins and destinations of trips requested by users are represented in the proposed model by pairs of nodes on the graph, $G(t)$. Consequently, $N_R(t)$ is the set of nodes in $N(t)$ relative to requests from the car-sharees (i.e. pedestrians, passengers):

$$N_R(t)=\bigcup_{p=1}^{n}\left\{R_p^+,R_p^-\right\}\subseteq N(t) \quad [1.4]$$

– P_p is the number of people, U_p included, to transport from the origin R_p^+ to the destination $R_p^-\left(P_p\geq 1\right)$. By default, this parameter equals 1, considering U_p as a single passenger.

–$[d_p,a_p]$ designates a limited time interval that defines the overall length of the maximum route tolerated by the user [DUM 91], where d_p is the earliest possible start time for the user, p, and a_p is the latest tolerated arrival time. Otherwise, failing to have specified preferences, the user will be assigned default values. These values correspond to the system time relative to the time of transmission of the request for the earliest departure time and the Estimated Time of Arrival (ETA) for the latest arrival time:

$$ETA(R_p^+, R_p^-, t) = \rho_{R_p^+, R_p^-}(t) + \tau(R_p^+, R_p^-) \qquad [1.5]$$

This time is calculated based on the two following functions:

1. A weighting function calculates the journey time $(\rho_{O,D}(t))$ required to travel a set route from an origin O to a destination D. In this case, the graph $G(t)$ is said to be "valued", allowing the assignment of a value (i.e. a weight) to each arc of the graph. The weight of an arc (O,D) such that $(O,D) \in A(t)$ equals $(\rho_{O,D}(t))$ is calculated as a function of several factors:

$$\rho O, D(t) = \text{Distance } (O,D) * \frac{1}{AJS(O,D,t)} \; with \; (O,D) \in A(t) \qquad [1.6]$$

One must note that in the context of urban and pre-urban transport, the weightings of the arcs cannot be negative [FEK 10]. Thus, $\forall (O,D) \in A(t), \rho_{O,D}(t) \geq 0$.

The weighting of arcs of the graph is calculated based on:

– the function that calculates the distance between two geographical points represented by their GPS coordinates (i.e. longitude, latitude). In this formula, the spherical law of cosine is taken into account and is as follows:

$$\text{Distance } (O,D) = \text{acos}\begin{pmatrix} \sin(\text{latitude }_O) * \sin(\text{latitude }_D) + \cos(\text{latitude }_O) * \\ \cos(\text{latitude }_D) * \cos(\text{longitude }_D - \text{longitude }_O) \end{pmatrix} * R$$

[1.7]

– the AJS parameter, which indicates the Average Journey Speed on a specific route at a specific time t. This parameter varies depending on several factors, amongst others the nature of the road, weather conditions, rush hour, etc.

2. The $\tau(O,D)$ function calculates the maximum possible delay on a route $[O,D]$. It is not considered a route segment unless there is an arc $(O,D) \in A(t)$ connecting both ends in the direction indicated. The rate of tolerable delay is calculated as a function of the distance between the origin and destination nodes of the route proportional to a Tolerable Delay Threshold (*TDT*), configurable by the system administrator:

$$\tau(O,D) = \text{Distance } (O,D) * \frac{\alpha}{\beta} \qquad [1.8]$$

TDT represents the value of tolerable delay (α) in minutes for a ride of (β) kilometers, to calculate the latest arrival time at the end of the route in question.

1.3.3. *Matrix structure to collect information on requests*

In what follows, a matrix structure is proposed for the collection and representation of information:

$$\begin{array}{c} \\ D(t) = \begin{array}{c} 1 \\ \vdots \\ i \\ \vdots \\ j \\ \vdots \\ k \end{array} \end{array} \begin{array}{c} \begin{array}{ccccccc} 1 & \cdots & i & \cdots & j & \cdots & k \end{array} \\ \begin{pmatrix} \phi & & \cdots & & \cdots & & \cdots \\ & \ddots & & & & & \\ \cdots & & \phi & & D_{i,j} & & \cdots \\ & \cdots & & \ddots & & \cdots & \\ \vdots & & D_{j,i} & & \phi & & \vdots \\ \cdots & & \cdots & & \cdots & \ddots & \cdots \\ \cdots & & \cdots & & \cdots & & \phi \end{pmatrix} \end{array}$$

Each element $D_{i,j}$ with $1 \le i, j \le k$ of the matrix $D(t)$ is a model of the existing demand on a set route $[a,b]$ starting from origin a going to destination b. The indices of the matrix refer to the nodes of the network. A list of variable size is used within each element of the matrix to represent the set of requests relative to the route represented by the indices of the element. In these elements, we find users who have requested to travel this route meaning that users whose entire requested trip $[R_\rho^+, R_\rho^-]$ or portion of the trip concerns the route $[a,b]$. These users are represented with all the specifics of travel required on the section in question (i.e. the preferred time interval and the number of passengers).

$$D_{i,j} = \bigcup_{p=1}^{n} \left\{ \left(U_p, P_p, \left[d_p, a_p \right] \right) \right\} \text{ such that } [a,b] \in \left[R_p^+, R_p^- \right] \quad [1.9]$$

We denote $O(t) = \bigcup_{j=1}^{m} \left\{ O_j(t) \right\}$ as the set of offers of m vehicles proposed for travel at time t.

$V(t) = \bigcup_{j=1}^{m} \left\{ V_j(t) \right\}$ is the fleet of vehicles having offered a bid for a specific trip.

To any specified car-sharing trips offered corresponds a couple of nodes in $N(t)$ representing its starting point (i.e. origin (O_j^+)) and destination (i.e. destination (O_j^-)):

$$N_o(t) = \bigcup_{j=1}^{m} \left\{ O_j^+, O_j^- \right\} \subseteq N(t) \quad [1.10]$$

$N_o(t)$ denotes the set of origin and destination nodes indicated by drivers of vehicles submitting car-pooling offers at time t.

Structure of a car-pooling offer

Offers of vehicles are modeled by our system in a similar manner to car-pooling requests previously described, as in the equation below:

$$O_j(t) = \left\{ O_j^+, O_j^-, l_j \begin{pmatrix} ID_1 \\ \cdots \\ ID_z \end{pmatrix}, \left[d_j, a_j \right] \right\} \qquad [1.11]$$

$O_j(t)$: offer of journey relative to car V_j

To submit an offer, all drivers must specify:

– $\left(O_j^+\right)$ as the starting point of their journey;

– $\left(O_j^-\right)$ as the destination;

– l_j represents the number of spaces available in the vehicle;

– $[d_j.a_j]$ defines the travel preferences of the car driver;

– the vector $ID_j = \begin{pmatrix} ID_1 \\ \vdots \\ ID_z \end{pmatrix} = \left(ID_{l,j}, \ldots, ID_{q,j}, \ldots, ID_{z,j}\right)^T$ indicates the set of Intermediate Destinations (*ID*) through which vehicle V_j must pass, from which time taken, *t*, to reach its final destination O_j^- is considered;

– we define $N_{ID}(t)$ as being the union of each one's intermediate destinations:

$$N_{ID}(t) = \bigcup_{j=1}^{m} \bigcup_{q=1}^{z} ID_{q,j} \subseteq N(t) \qquad [1.12]$$

We then define $N(t)$ as follows:

$$N(t) = N_R(t) \bigcup N_O(t) \bigcup N_{ID}(t) \quad [1.13]$$

1.3.4. *Matrix representation for modeling car-pooling offers*

In order to take into account all existing car-pooling offers, we propose to model the offers of all vehicles of the fleet in an Origin Destination matrix $I(t) = (r_{o,d})_{l \leq o,d \leq f}$.

The indices o and d of elements $I(t)$ here refer to the nodes of origin and destination of a route $\left[a,b\right]$.

Each index represents a node that can be either origin or destination or both at the same time, representing in turn different segments of the route (e.g. $\left[a,b\right]$ or $\left[b,a\right]$) depending on the offer being considered.

This matrix representation is based on a decomposition of the set of overall itineraries $\left[a,b\right]$ on elementary roads such that $\left(a,b\right) \in A(t)$. This decomposition is a figure of disassembly of vehicle routes based on their origins, destinations and visited points:

$$\begin{array}{l} a \in \{N_O(t) \cup N_{ID}(t)\} \text{ and } b \in \{N_O(t) \cup N_{ID}(t)\} \\ \forall o \in \{1...f\} \text{ and } \forall d \in \{1...f\} \end{array} \quad [1.14]$$

The indices o and d of matrix $I(t)$ refer to the set of nodes where $f = \left|N_o(t) \bigcup N_{ID}(t)\right| = Card(N_o(t) \bigcup N_{ID}(t))$, wherefrom $f \leq 2m + Z$, with $Z = \left|N_{ID}(t)\right| = Card(N_{ID}(t))$. The set of offers are thus reduced to a matrix representation of the sets of routes taken, considering down to the smallest part of route $\left[a,b\right]$ defined by an origin a to a destination b:

$$I(t) = \begin{array}{c} \\ 1 \\ \vdots \\ o \\ \vdots \\ d \\ \vdots \\ f \end{array} \begin{array}{c} \begin{array}{ccccccc} 1 & \cdots & o & \cdots & d & \cdots & f \end{array} \\ \left(\begin{array}{ccccccc} \phi & & \cdots & & \cdots & & \\ & \ddots & & & & \cdots & \\ \cdots & & \phi & & r_{o,d} & & \phi \\ & \cdots & & \ddots & & \cdots & \\ X & & r_{d,o} & & \phi & & \\ \cdots & & & & & \ddots & \\ & \cdots & & X & & & \phi \end{array} \right) \end{array}$$

Each element $r_{o,d}$ of the matrix represents the supply side of the route section represented by $[a,b]$. The indices of the matrix, being represented by the overall set of origins, destinations and intersection points of all cars, allow route sections to be represented in the same matrix without being used or even allowing passage in the direction indicated. The matrix elements can, as the case may be, have one of the following values:

$$r_{o,d} = \begin{cases} X: \text{if according to the highway code, } [a,b] \text{ cannot be inserted in the direction indicated,} \\ *: \textit{if } [a,b] \text{ is practically doable but no offer exists (i.e.no vehicle covers this area)} \\ \bigcup_{j=1}^{m} \left\{ \left(Vj, l_{a,b}^{j}, \left[d_a^j, a_b^j \right] \right) \right\} \text{such that} [a,b] \in \left[O_j^+, O_j^- \right] : \text{The set of vehicles intersecting route} \\ [a,b] \text{ with, for each ,different displacement parameters} \end{cases}$$

$r_{o,d}$ contains the list of parameters specific to each vehicle serving the section of journey in question, namely:

– $l_{a,b}^{j}$ is the number of spaces available in vehicle V_j on route $[a,b]$;

– $\left[d_a^j, a_b^j\right]$ indicates for all vehicles V_j serving $\left[a,b\right]$, the earliest starting time d_a^j from the starting point a of $\left[a,b\right]$ and the latest arrival time a_b^j at the end point b.

The elements of the matrix $I(t)$ are relative to route sections being used or not by one or more vehicles. These routes are thus obtained by a decomposition of offered trips.

Any vehicle V_j of the fleet is associated with an overall itinerary IT_{V_j}, constructed in real time at the considered time t. Subsequently, $\forall j \in \{1..m\}$, is defined as a string of ordered nodes ($v_0, v_1, \ldots v_n$), with $n = z + 2$ and $z = \left|ID_j\right| = Card\left(ID_j\right)$:

$$IT_{V_j}(t) = (v_0, v_1, \ldots, v_n) \text{with} \begin{cases} v_k \in N(t), \forall k \in [0,n] \\ v_0 = O_j^+, v_n = O_j^-, \bigcup_{k=1}^{n-1} v_k = ID_j(t) \\ (v_k, v_{k+1}) \in A(t), \forall v_k, v_k + 1 \in IT_{V_j}(t); k \in \{0..n-1\} \end{cases}$$

Every two successive nodes are connected together by an arc in $A(t)$ and thus the overall itinerary of the car is defined as the adjacent routes starting from the origin of the journey and going towards the defined end point. In the following formula (equation [1.15]), an itinerary IT_{V_j} is defined as a junction of n – 1 Partial Itinerary (PI) so long as it consists of n successive nodes:

$$IT_{V_j}(t) = \left(PI_1^j \oplus \ldots \oplus PI_{n-1}^j\right) = \left(\left[O_j^+, ID_1\right], \ldots, \left[ID_{q-1}, ID_q\right], \ldots, \left[ID_z, O_j^-\right]\right) \tag{1.15}$$

PI_s^j represents the Partial Itinerary number s in the overall itinerary $IT_{V_j}(t)$ of V_j; $\oplus$ is used here as a juxtaposition operator to show the succession of route sections designated by *PI* in a fixed order such as to form the overall itinerary. Each *PI* is modeled by its corresponding element in the matrix $o \in N_h^+(t)$; it has its own properties and can include one or more offers.

In addition, for each offer on a particular *PI*, the parameters specific to it are automatically calculated by the system. Indeed, the number of spaces available is updated at each pick up and drop off of a passenger. It is also possible to calculate the time interval extremities for the start and end on any part of route $\left[o,d\right]$, as soon as the necessary information on the time course of the preceding route segment is obtained. By "necessary information", we specifically mean the earliest departure time and the latest arrival time of vehicle V_j on the route section $\left[h,o\right]$,which directly precedes $\left[o,d\right]$. $\left[h,o\right]$ precedes $\left[o,d\right]$ if, $o \in N_h^+(t)$, $d \in N_o^+(t)$, $(h,o) \in A(t)$, $(o,d) \in A(t)$ and: $\left[\left[h,o\right],\left[o,d\right]\right] \subseteq IT_{V_j}(t)$

$$\begin{cases} d_o^j(t) = d_h^j(t) + \rho_{h,o}(t) = \text{the earliest arrival time at o} \\ a_d^j(t) = d_o^j(t) + \rho_{o,d}(t) + \text{Distance}(O_j^+, d) * \dfrac{\left(a_{O_j^-} - d_{O_j^-}\right)}{\text{Distance}\left(O_j^+, O_j^-\right)} \end{cases}$$

$\tau(o,d) = \text{Distance}\left(O_j^+, d\right) * \dfrac{\left(a_{O_j^-} - d_{O_j^-}\right)}{\text{Distance}\left(O_j^+, O_j^-\right)}$ is the possible delay in arriving at *d* from *o*. This delay is estimated based on the delay tolerated by user $\left(a_{O_j^-} - d_{O_j^-}\right)$ across the entire trip. We then give $\alpha = a_{O_j^-} - d_{O_j^-}$ the value of the delay that the driver can tolerate for

for their overall journey and $\beta = \text{Distance}(O_j^+, O_j^-)$. $a_{O_j^-}$ and $d_{O_j^-}$ respectively express the latest arrival time set by the user and the earliest time of departure (arrival) to O_j^- .

1.3.5. *Modeling constraints of vehicles’ allocation to users*

In what follows, we consider an approach that suggests considering the individual travel preferences of the car-pool applicant as well as the car-pool drivers offering the trips, whose parameters are set in advance but still somewhat flexible. It is therefore important to consider a certain degree of flexibility in relation to the construction of itineraries in order to meet travel requirements.

This results in the creation of an itinerary composed of several sub-itineraries, to which different vehicles are assigned each time. Figure 1.1 shows an example of a route composition solution for the user U_i from the available fleet where several vehicles can be considered if a single vehicle cannot alone meet the overall request.

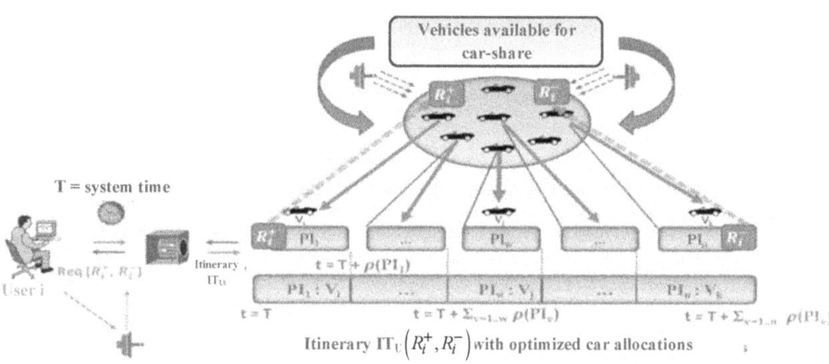

Figure 1.1. *Composition of solutions based on allocation of available supply*

Multiple offers on the same route allow us to choose the one that best minimizes the overall length of the trip: an optimization criterion chosen at system level. Thus, the car that best optimizes the overall travel time on each section in question is ultimately selected to build the final solution:

$$S_{U_i}(t) = (S_{1,i}\ o \cdots o\ S_{w,i}\ o \cdots o\ S_{n,i}) \qquad [1.16]$$

where $S_{U_i}(t)$ is a comprehensive optimized solution provided to the user, U_i, at the end of the processing of request $S_{w,i}$. This solution is, as equation [1.16] shows, reconstructed from a sequence of several partial solutions o_j that are relative to each part of the itinerary to be taken by U_i in order to achieve their goal. Here, "o" is a combination operator of partial solutions $R_i(t)$, indicating compliance with the order and direction in which the solutions are displayed, and with the image of the sequence of partial itineraries $\left(PI_{w,i}\right)$ relative to them. Each partial solution $S_{w,i}$, $\forall w \in \{l..n\}$, is then modeled as a data structure containing the necessary information for the route section in question. Aside from the pick-up $\left(PI_{w,i}\right)$ and drop-off $\left(PI^{-}_{w,i}\right)$, this structure provides data essential to the solution, namely, details about the vehicle in question (V_j), the earliest departure time and latest end time on the extremities of the itinerary:

$$S_{w,i} = \left(PI^{+}_{w,i}, PI^{-}_{w,i}, V_j, d_{PI^{+}_{w,i}}, a_{PI^{-}_{w,i}} \right) \qquad [1.17]$$

The solution itinerary provided to the user U_i is then formulated as follows:

$$IT_{U_i}(t) = \left(PI^{v}_{l,i} \oplus \cdots \oplus PI^{j}_{w,i} \oplus \cdots \oplus PI^{k}_{n,i} \right) \qquad [1.18]$$

Where $PI^{j}_{w,i}$ is the wth Partial Itinerary served by vehicle V_j in the overall route defined as a result of processing the request of user *i* (i.e. $R_i(t)$). Here, *j* refers to the car that has the optimum journey time out of V_w cars on the route section in question, $PI^{j}_{w,i}$. V_w is the set of vehicles whose respective itineraries include the path $PI^{j}_{w,i}$ in the direction indicated, whilst satisfying the correspondence constraints and thus forming a set of "*feasible*" solutions for the considered route section.

Thus, partial itineraries for common routes are processed firstly in such a way as to extract the set of feasible solutions, meaning the cars that intersect the corresponding routes in the appropriate direction, in the desired time period and providing a sufficient number of spaces. Next, the car with optimal travel time is selected.

The algorithms we propose are executed following two main steps: firstly, we start by finding the set of feasible solutions V_w and secondly, the optimal solution among them.

To be considered feasible, a proposal for an offer O_j on a stretch of route $PI_{w,i}$ must satisfy a set of constraints that match the following criteria:

– $d_{PI^j_{w,i}} \geq a_{PI^v_{w-1,i}} - \tau\left(PI^+_{w,i}, PI^-_{w,i}\right), \forall w \in \{2..n\}$: a vehicle V_j can only be considered in the allocation process if it satisfies the first corresponding constraint (i.e. matching $PI_{w,i} \in IT_{V_j}$) between its overall itinerary and $PI_{w,i}$. This is reflected by the fact that V_j must travel its departure route section $l^j_W \geq P_i$ to its final destination $\left(PI^-_{w,i}\right)$ in the appropriate direction.

– $l^j_W \geq P_i$: the vehicle V_j is only a feasible solution if it provides a number of vacancies greater than or equal to the number of passengers seeking to travel on the route $PI_{w,i}$ as indicated by the request inserted by user U_i.

– $d_{PI^j_{1,i}} \geq d_{U_i}$: the earliest start time specified by the driver of V_j on the first portion of route $PI_{1,i}$ in IT_{U_i} must match that specified by U_i. The route section $PI_{1,i}$ goes from the origin of travel of passenger U_i, thus forming the initial part of the overall itinerary.

– $d_{PI^j_{w,i}} \geq a_{PI^v_{w-1,i}} - \tau\left(PI^+_{w,i}, PI^-_{w,i}\right), \forall w \in \{2..n\}$: moreover, for the rest of the sections that can be combined to make the overall itinerary, a car is only considered feasible on a part of route $\left(PI_{w,i}\right)$ if its earliest departure time from $\left(PI^+_{w,i}\right)$ is greater than or equal to the earliest arrival time at the end of the route that directly precedes it $\left(PI_{w-1,i}\right)$, regardless of which vehicle v is assigned to it (bearing in mind that $PI^+_{w,i} = PI^-_{w-1,i}$).

– $a_{PI^j_{n,i}} \leq a_{U_i}$: this final constraint particularly concerns the latest arrival time by vehicle V_j that can be assigned to the last stretch $\left(PI_{n,i}\right)$ of user i's itinerary. The proposed system must meet the margin of delay tolerated by the user U_i. A vehicle V_j can only be assigned if it reaches the end of the course at a time less than or equal to the latest arrival time specified by U_i. Consequently, the vehicle V_j must reach the final end of the last section of route $\left(PI^-_{n-1,i} = d^-_{U_i}\right)$ of itinerary $IT_{U_i}(t)$ at a_{U_i} at the latest.

The verification of these constraints is done for all car-pooling offers issued at a time t in order to extract those that are likely to be considered as potential solutions to the request being processed. The set of feasible solutions on a part of the constructed itinerary is progressively obtained. This set, called V_w, refers to all vehicles serving the route labeled w of the itinerary of user U_i, while satisfying all the constraints listed above.

Since we mainly focus on our approach of optimization solutions provided to the user, a single solution is chosen for each route w in such a way as to optimize the itinerary. The criteria we have set refer to the Global Trip Duration (TD) which is calculable in terms of waiting time and travel time. These are, in turn, calculated using the weighting function (ρ) defined above. A fitness function allows aggregation of waiting times ($W_{j,w}$) and travel ($T_{j,w}$) on a route section ($PI_{w,i}$), thus allowing the calculation of travel time for any vehicle $V_j \in V_w$:

$$Fitness = TD^j_{PI_{w,i}} = \omega_{1_i} W_{j,w} + \omega_{2_i} T_{j,w} \text{with } V_j \in V_w \qquad [1.19]$$

ω_{1_i} and ω_{2_i} are the aggregation parameters with a weighting of $W_{j,w}$ and $T_{j,w}$ respectively. They are set by the user and indicate their travel preferences by favoring optimization relative to waiting rather than travel time or spending less time in a car, even if it affects the waiting time.

The final solution $TD^*_{PI_{w,i}}$ is that which minimizes the fitness function on the section w, based on the objective function defined as the following:

$$TD^*_{PI_{w,i}} = Min_{V_j \in V_w} (TD^j_{PI_{w,i}} = \omega_{1_i} W_{j,w} + \omega_{2_i} T_{j,w}) \qquad [1.20]$$

Based on the overall objective function thus defined and jointly with the principle of recomposing a complete solution from several partial solutions, the solution that optimizes the overall itinerary $\left(IT_{U_i}\right)$ is the one that optimizes the combination of all sections $w \in \{l..n\}$ of which the intersection constitutes the route $\left[R_i^+, R_i^-\right]$.

Given that our initial objective is to optimize the length of the entire journey carried out by U_i, the final solution is therefore to consider within each section $w \in IT_{U_i}$ the vehicle minimizing travel time. The sum of the optimized lengths on all sections of the itinerary constituting the full itinerary consequently leads to the minimization of total time spent traveling on IT_{U_i}:

$$TD^*_{U_i} = \sum_{w=1}^{n} Min_{V_j \in V_w} \left(TD^j_{PI_{w,i}}\right) = \sum_{w=1}^{n} Min_{V_j \in V_w} \left(w_{1_i} W_{j,w} + w_{2_i} T_{j,w}\right)$$

[1.21]

1.3.6. *Geographical network subdivision served and implementation of a physical distributed dynamic architecture*

In order to alleviate the impact of heavy processes on the quality of service provided and response time, we propose to develop a platform that promotes the decomposition of the initial overall process into several less complex tasks. These tasks will be carried out in a decentralized manner and in parallel to reduce the complexity of processes performed.

To do this, a status representation of the current geographic network served is established as a distributed dynamic graph, taking inspiration from some existing works using the same principle [FEK 10; KAM 07; HIZ 08].

a) Decomposition areas

For the decomposition of the network served, considering an effective method to encompass all available data while avoiding duplication of information overrides any other objective. We propose to achieve this by a principle of subdivision to create a set of distinct geographical areas whose main features are:

– All areas created represent geographical areas of similar shapes and sizes.

– Each area has its own perimeter that encompasses an information set relating to requests and/or offers of a car-sharer. Thus, each zone corresponds to a software component, more or less independent, locally processing the data included in each area. A set of autonomous (or semi-autonomous) components is created for established areas. They run in parallel so as to optimize the overall system processing.

– Two or more zones may have common areas of intersection, representing a single point or several points as they may represent an empty geographical area.

– Two or more zones can intersect and share data which could help proper implementation of the system; the data redundancy within zones is useful because each component created needs comprehensive knowledge of the system in order to best address the user requests located in specific areas. However, it can conflict when two or more areas share data on the origins of user requests.

– In what follows, we focus particularly on methods for the classification of data according to one or more matching criteria for the creation of scatter points. The idea behind these methods, and especially in "*unsupervised classification*", is to create a set of *k* classes, where each class corresponds to a set of individuals. Areas of high density are then responsible for creating "clusters" with which class labels are associated. Observations on the network are thus analyzed and studied based on well-defined criteria, in this case distance, in order to later be able to collect a set of neighboring classes in space.

– The network served that we model is reduced to a two-dimensional space where the information that we have is presented as a homogeneous dataset. These data relate to geographical points represented by their GPS coordinates (i.e. longitude and latitude). Taken together, these data, which constitute the observations of the system, can be classified in such a way that the dominant character of separation is the distance between each two nodes of a population belonging to the same cluster. A real underlying class hence occupies a limited time–space region of the map representing the network area served. Each class is formed relative to a set of nearby nodes observed by the previously detailed graphical modeling, which may be decomposed and thus give rise to a multitude of clusters.

– As we have previously observed, two or more zones can overlap and share one or more individuals (i.e. nodes) with different degrees of membership and whose sum is equal to one [AMM 07]. Based on this principle, we are then placed in the context of an unsupervised blurry classification [ALL 06] where the number of groups is unknown and must be defined by the decomposition algorithm. Also, being placed in the context of a dynamic car-pooling service where supply and demand can occur at any time, it is imperative to consider the evolution of the number of individuals for continuous updating of classes thus established. The number of clusters may change during the evolution of processes and hence throughout the allocation process or the decomposition process.

– Each zone is associated with a region of "*similar*" observations according to the established criteria (i.e. close in space according to an initial pre-determined limiting distance). Based on what has preceded it, an area contains the following parameters:

- cardinal (N): the number of individuals;

- center (Ci): the center of area Zi;

- radius (r_i): the radius of the area Z_i . This radius is the same for all established areas.

– From the set of determined neighbors at each iteration of the subdivision algorithm detailed later, the properties of each class are created and hence defined, essentially its center and radius. From these parameters, new populations can be created determining the breadth of knowledge for the identified areas ("classes") [ALA 06]. The center of the area is calculated so that it is the equidistant point (i.e. center of gravity) between the furthest points of the set being considered. Several scenarios can occur and should be considered as long as the set of determined neighbors contains one, two, three or more nodes:

- the determined set contains a single node removed from the other points in the network at a strictly greater distance than the predetermined diameter of the system;

- if the neighboring points are two in number and no node belongs to the perimeter of the area including that pair of nodes, the center of the area that includes them is calculated as the midpoint of the line connecting the two points in question;

- the area in question is subsequently established based on the equidistant point calculated from the two given nodes and its, in theory, pristine neighborhood. However, the knowledge space of an area may evolve to include other nodes related to new offers or requests for car-pooling;

- if the set of neighbors has a cardinal of three, then the center of the area that contains all of these nodes is the center of gravity of the triangle whose vertices are represented by points of this set;

- if the set of neighbors is of size three or more (Cardinal (N)≥3), then the center of the area that contains all of these nodes is the center of gravity of the triangle whose vertices are represented by points of this set. In fact, it is the center of gravity identified as the isobarycenter of these three vertices and the center of mass of the interior of the triangle. This point is located two-thirds of the way along from each center starting from the vertex.

b) Decomposition principle

Step 1: Creation of Primary Zones

Algorithm 1.1 describes the software steps followed by our system in order to create the Primary Zone. These steps rather refer to a set of iterations during which the sets of neighbors are defined and their respective centroids are identified, on the basis of what the areas are established on.

Algorithm 1.1. Creation of Primary Zones (CPZ)

Data: $SN = N_R(t)$: set of origins and destinations nodes of requests, Diameter

Result: PZ: Set of Primary Zones

Initialization: Number k = 0, Distance = 0, VN: Set of nodes visited = ϕ, N: Table of set of neighbors = ϕ

While $VN \neq SN$ **Do**

Random selection of a node n of $\{SN - VN\}$

$N[k] \leftarrow N[k] \bigcup n$

$VN \leftarrow VN \bigcup n$

For All node n_1 **such that** $n_1 \in \{SN - VN\}$ **Do**

For All node n_2 **such that** $n_2 \in N[k]$ **Do**

$Distance = Distance\ (n_1, n_2)$

If $Distance \leq Diameter$ **Then**

$N[k] \leftarrow N[k] \bigcup n1$

$VN \leftarrow VN \bigcup n1$

End If

End For

End For

$k \leftarrow k+1$

End While

For All number i **such that** $i \in \{0 \cdot\cdot k\}$ **Do**

$C_i \leftarrow Center(N[i])$

$PZ_i \leftarrow Establish\_Primary\_Zone(C_i, Diameter)$

$PZ \leftarrow PZ \bigcup PZ_i$

End For

Return PZ

Step 2: Subdivision of the network

In this second step of subdivision of the network, all information (*node*) not processed in the first step is taken into account. This information is considered with the aim of finding allocations for areas already established or creating new ones based on the same neighborhood principle.

These areas are said to be secondary areas or Intermediate Zones (*ZI*), since they only enter the allocation process if it is necessary. The overall set Z includes all the areas created, whether primary or secondary, that are likely to be questioned, updated or perhaps even eliminated, depending on the flow of information received. This is a view of optimal processing, avoiding duplication of information and/or processing, while covering the search space in question.

Algorithm 1.2. Creation of Intermediate Zones (CIZ)

Data: $SN = N_O(t) \bigcup N_{ID}(t)$: set of origin, destination and intersection point nodes of vehicles, number k

Result: IZ: Set of Intermediate Zones

Initialization: Number m = k, Distance = 0, Diameter, VN: List of nodes visited = ϕ, N: List of neighbors = ϕ, Boolaen*Allocated* = *False*

While $VN \neq SN$ **Do**

Random selection of a node n of$\{SN - VN\}$

$VN \leftarrow VN \bigcup n$

For All number j **such that** $j \in \{0 \cdot\cdot m\}$ **Do**

If $j \leq k$ Then

$C_j \leftarrow Center(Z_j)$

Distance $\leftarrow$ Distance(n, C_j)

Otherwise

Distance $\leftarrow$ Greatest$(N[j])$

End If

If $Distance \leq Diameter$

Then

$N[j] \leftarrow N[j] \bigcup n$

Allocated $\leftarrow$ True

End If

End For

If Allocated = False **Then**

$m \leftarrow m+1$

$N[m] \leftarrow N[m] \bigcup n$

End If

End While

For All i **such that** $i \in \{k \cdot\cdot m\}$ **Do**

$C_i \leftarrow Center(n[i])$

$IZ_i \leftarrow Establish\_Intermediate\_Zone(C_{i,} Diameter)$

$IZ \leftarrow IZ \bigcup IZ_i$

End For

Return IZ

1.4. ODCCA: an optimized dynamic car-pooling platform based on communicating agents

The necessity for the establishment of an effective process that combines the proposed "*physical*" architecture of the land is all the more imminent as it contributes to the effectiveness of processing, while following the evolution of the image of the car-pooling network. This contribution lies with the fact that establishing a decentralized and relocated request optimization process of several software entities works in parallel. These entities that we place in a multi-agent system context are also assimilated to *"distributed"* software agents, similar to the graphical modeling proposed in the previous section. Their main role is to carry out an optimized and efficient allocation of system data in the way that they are distributed in the geographical network served.

1.4.1. *Multi-agent concept for a distributed car-pooling system*

The modeling of a geographical network as a Distributed Dynamic Graph is conducive to the deployment of a software architecture composed of a multitude of autonomous entities operating in parallel and promoting the reduction of processes by reducing the complexity of the problem. In what follows, we propose to establish a multi-agent system called ODCCA, which stands for Optimized Dynamic Car-pooling based on Communicating Agents, which is provided with a multitude of features.

Functional analysis of the dynamic car-pooling problem

In what follows, five major steps (Figure 1.2) are required with the ultimate aim of ensuring a high quality service with a maximum rate of positive responses. The idea encompasses different preliminary steps preceding the application of algorithms for the allocation of vehicles to users.

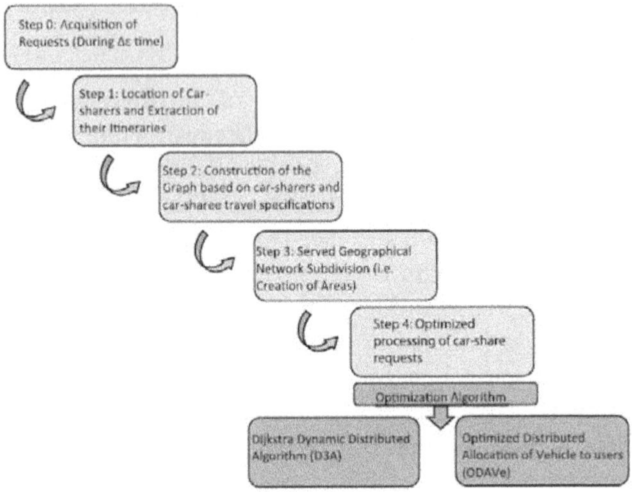

Figure1.2. *Features of the proposed system*

Step 1: Acquisition of requests

This involves the ability to simultaneously acquire specifications for car-sharers issuing simultaneous or near-simultaneous requests. A new parameter Δε has been introduced; it consists of a negligible lag time during which the system will wait for any car-pooling requests before starting the allocation search process.

Two pivotal functions govern the parallel processing of data on pedestrians and their travel requirements.

The first relates to the acquisition of requests issued during the time lag indicated (i.e. Δε).

The second relates to the organization of the set of requests received in order to highlight similarities and promote parallel processing (Figure 1.3).

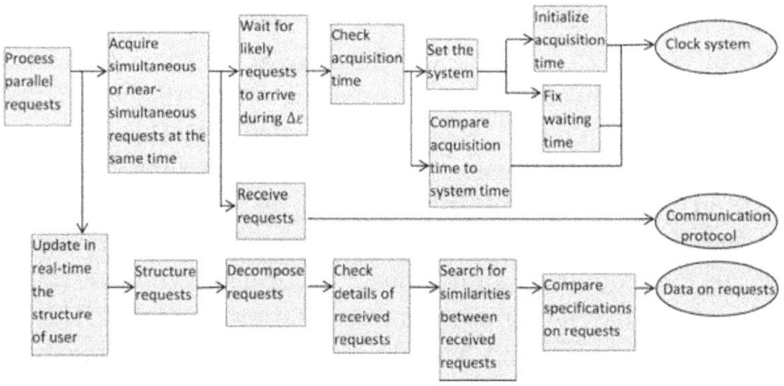

Figure 1.3. *Reception and structure of quasi-simultaneous requests*

Step 2: Extraction of car-pooling offers

The proposed system must process real-time offers of journeys by continually updating the search space to include the new features of the offer in question. A continuous and dynamic update aims to integrate any newly received offer in order to consider it in the allocation process and perhaps offer it as a possible solution, if it matches.

Figure 1.4 illustrates the operation of the system for the identification of car-pooling offers.

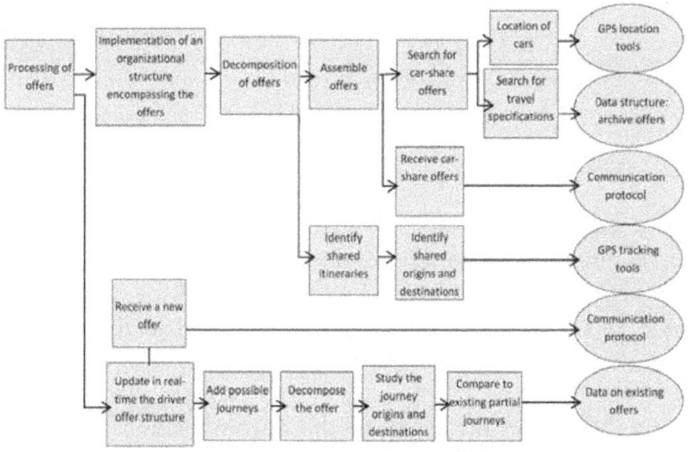

Figure 1.4. *Extraction and organization of car-pooling offers*

Step 3: Construction of the graph

From the two steps above, the matrix representations have resulted in the emergence of the concept of graphics based on graph theory in order to illustrate a real view of the served network. This step involves the establishment of a fusion graph where a superimposition of two planes in space is performed; the first represents the requests of pedestrians and the second represents driver offers. A continuous real-time update of the graph must be established and eventual modifications may occur when an external event occurs. The FAST diagram shown in Figure 1.5 shows the process of deploying different system input data in graph form.

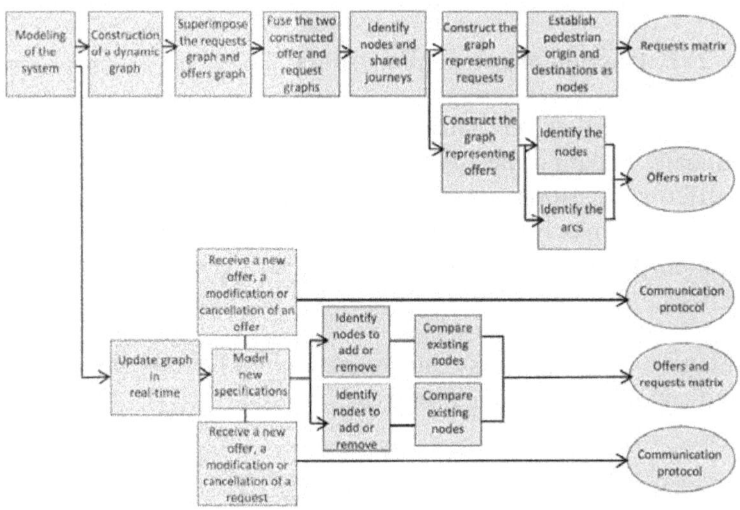

Figure 1.5. *Construction of a representative model of the system*

Step 4: "Divide and conquer"

In this section, we propose subdividing the geographical network served to highlight areas of greater or lesser density. Consequently, the process of finding optimized allocations will be decomposed and relocated in the extracted areas and is thus led to the image of the network after subdivision.

Figure 1.6 shows the result of the decomposition principle stated above.

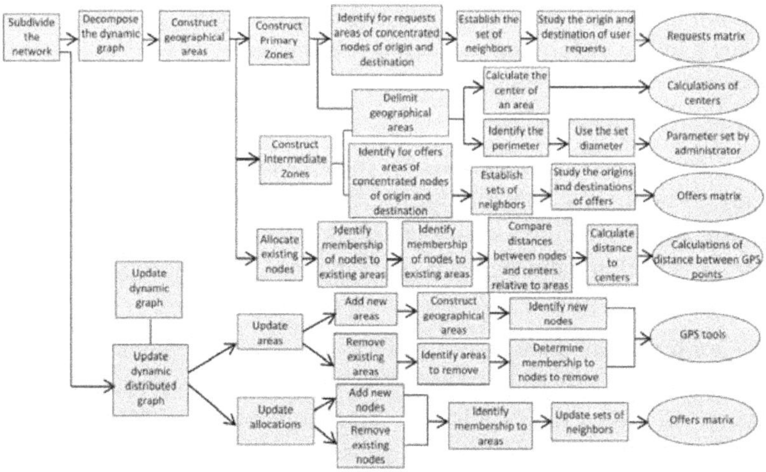

Figure 1.6. *Decomposition of the network*

ODCCA: A multi-agent architecture

Given the exponential complexity of the problem of optimized dynamic car-pooling, a decentralization and division of tasks into a set of software components is required. These components are presented as autonomous intelligent entities running in parallel in a rapidly changing and unstable environment. A multi-agent architecture is therefore proposed in Figure 1.7.

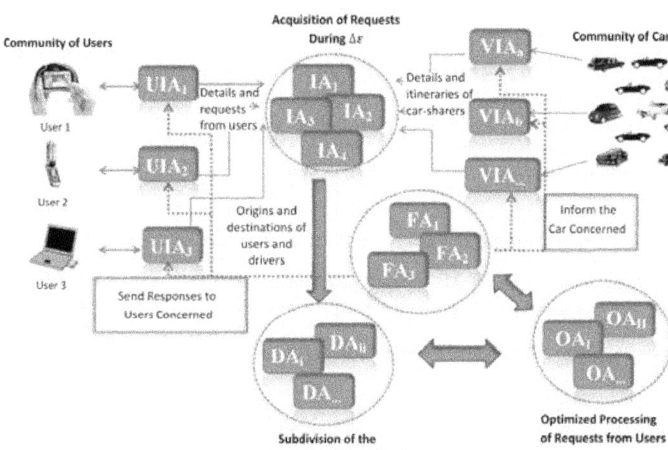

Figure 1.7. *ODCCA: a system based on Communicating Agents*

Each agent has its own action space and features. In what follows, we present the different agents currently in place, as well as details on their operation and their contributions to the optimized allocation process of vehicles to users.

– **User Interface Agent (UIA):** immediately from their connection to the system, any user of the service is supported by a UIA. The features of a UIA are summarized as receiving a request for car-pooling, checking the coordination and defining the default values specified by the user.

– **Vehicle Interface Agent (VIA):** this is to provide, at the core of the system, an interface through which it can communicate with drivers of cars. In parallel to communication with pedestrians, potential passengers, the system must be able to interact with car-sharers. These exchanges will be at the service of the coordination between drivers and passengers to be able to organize their journeys and automatically synchronize pick-ups and drop-offs. A (VIA) agent will also collect the necessary information about their offers and check the settings to transmit information directly to the information agent (IA) if the process is already underway.

– **Information Agent (IA)**: the function of this agent is to collect data necessary for the proper functioning of the system, bearing in mind that any existing information may contribute to the description of the state of the environment and organize them in a suitable data structure.

– **Decomposing Agent (DA)**: this agent is responsible for the process of subdividing the served network. The pre-defined decomposition algorithms are driven by the agent for the establishing sets of neighboring nodes formed on the basis of data received from the IA on offered and requested routes. The DA extracts "*communities*" of nodes in proximity. These help define zones, whether they are primary or secondary, and their geographical stretch due to the calculation procedure of the centers and the preset diameter.

– **Optimizing Agent (OA)**: two types of optimizing agent are implemented in our system. Each type depends on a well-defined optimization algorithm which is located in the core of the corresponding agent. Moreover, both share optimization tasks on different instances initiated by the corresponding type. They perform their tasks locally in the area for which they are responsible then come together in a coalition in order to reconstruct complete and consistent solutions. The bodies in question, whether they belong to one or the other of these two types, converge in a way that their behaviors fall within the context of distributed optimization.

– **Fusion Agent (FA)**: this agent is instantiated as soon as a set of solutions is generated by the system. It is responsible for identifying the user involved in each solution received, identifying the nature of the response (empty or invalid) and acting accordingly. It is also equipped with the ability to be the middle agent between the UIA and the VIA to validate, or not, a trip for the users concerned.

Entities established in our system must allow full synchronization without conflict and maintain an ambient intelligence throughout the system. For this, the establishment of communication channels between the agents is required, which is where the selection of an appropriate modeling language for this context, the UML, comes from.

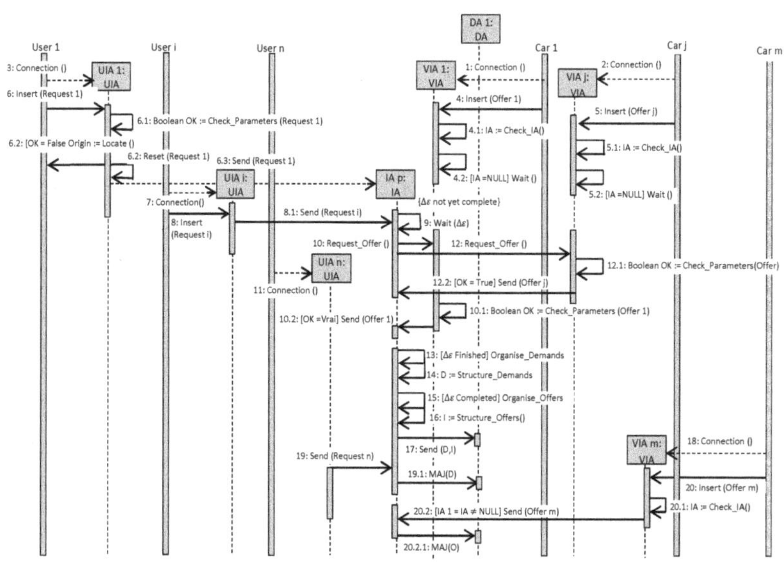

Figure 1.8. *Collection and organization of data on the supply and demand of car-pooling*

Figure 1.8 shows the collection and organization of data on carpooling supply and demand. The purpose of this coalition is to gather information and to structure it so that they can be manipulated by the DA and OA, and thus pave the way for the allocation processing of vehicles to users.

1.5. Formal modeling: for an optimized and efficient allocation method

After extraction from space of possible solutions, the proposed system goes on to the second stage of its operation; the search for the best solution, given the optimization criteria set here, is established as the overall duration of the journey. A different approach is proposed in each case based on two principles of adopted resolution. The first is an adaptation of Dijkstra's algorithms and the second consists of an exact optimization method. Both approaches are detailed in this section.

1.5.1. *D3A: Dijkstra Dynamic Distributed Algorithm*

In this first approach, we are interested in finding optimized itineraries based on the principle of Dijkstra distributed algorithms and to distribute the execution primarily on *DA* and *OA*. This process corresponds to the execution of Algorithm 1.3 provided in the following:

Algorithm 1.3. Dijkstra Dynamic Distributed Algorithm (D3A)

Data : $G = (N, A)$: Graph modeling the network, R : Requests by users,

Result : OS : Optimized Solutions

Initialization :

Node $D \leftarrow NULL$

For All node n **such that** $n \in N$ **Do**

$Weight[n] \leftarrow \infty$

End For

For All request r_i **such that** $r_i \in R$ **Do**

$Weight\left[r_i^+\right] \leftarrow d_i$

End For

For All OA_i **Do**{only the OA_i on the PA containing sources are launched}

$Weight_{OA_i}, Car_{OA_i} \leftarrow A3D(G_i, TPM, LV, R_i)$ {Launch algorithm *D3A* on each OA_i }

End For

For All request r_i **such that** $r^i \in R$ **Do**

$D \leftarrow r_i^-$

While $\left(D \neq r_i^-\right)$ **Do**

Insert D at the start of SO_i

$Z_i \leftarrow Locate(D)$ {Determine the zone that contains D to consider results provided by *OA* concerned}

$OS_i\left[\Pr edecessor\left[D\right], D\right] \leftarrow Car_{OA_i}\left[\Pr edecessor\left[D\right], D\right]$

$Pl_{Car\left[\Pr edecessor\left[D\right], D\right]} \leftarrow Pl_{Car\left[\Pr edecessor\left[D\right], D\right]} - P_i$

$D \leftarrow \Pr edecessor\left[D\right]$

End while

End For

$OS \leftarrow OS \bigcup OS_i$

Send *SO* to *FA*

Algorithm 1.4 provides the detail of application of the Dijkstra algorithm locally to each zone by the responsible OA. This algorithm is performed in order to look for possible allocations, particularly those that optimize the waiting and travel times, thus promoting earliest arrival at the desired destination.

Algorithm 1.4: Dijkstra algorithm optimized for the allocation of vehicles

---

Data: $G_i = (N_i, A_i)$: Graph modeling of zone Z_i , TPM (average travel time function), R_i : Requests whose origins are in Z_i , LV: List of Vehicles.

Result: *Weight*: List of node weights, *Car*: List of cars affected by sections of itineraries

Initialization:

For All node n **such that** $n \in N_i$ **Do**

$Weight[n] \leftarrow \infty$

End For

For All request r_i **such that** $r_i \in R_i$ **Do**

$Weight[r_i^+] \leftarrow d_i$

End For

$Q \leftarrow N_i$

While $Q \neq \phi$ **Do**

$p \leftarrow \min(weight)$ {choose the node with smallest weight}

$Q \neq Q - \{p\}$

For All node q **such that** $(p, q) \in A_i$ **Do**{nodes of the set of

---

neighbors of p }

For All $o_j \in O$ **Do**

If $(Check\_Constraints(o_j, r_i))$ **Then**

$W_t \leftarrow w_1 * Distance(o_j^+, p) * TPM(o_j^+, p, d_j)$

If $(W_t + t - Weight[p] \succ 0)$ **Then**

$W_{t_j} \leftarrow W_t$

End If

End If

End For

$c \leftarrow \min(W_t)$ {Choose car with shortest waiting time}

$a \leftarrow Weight[p] + W_{t_c} + w_2 * Distance(p, q) * TPM(p, q, Weight[p] + I$

If $a \prec Weight[q]$ Then

$Weight[q] \leftarrow a$

$Predecessor[q] \leftarrow p$

$Car[p, q] \leftarrow c$

End If

End For

End while

Send *Weight* to adjacent areas

Return *Weight* and*Car*

1.5.2. *ODAVe: Optimized Distributed Allocation of Vehicle to users*

In order to improve system performance, this second approach has been suggested with the aim of supporting an optimized "*if possible*" allocation search principle, while focusing on collaborative work between different *OAs*.

Indeed, as shown in Figure 1.9, the idea is to find the nearest node served, decompose the request, solve the first stage of the request, send a message of type "Request/Response" to the OA responsible for the zone containing the nearest node and then repeat on the next closest set of nodes served until the solution is non-empty or until there are no more nodes to visit.

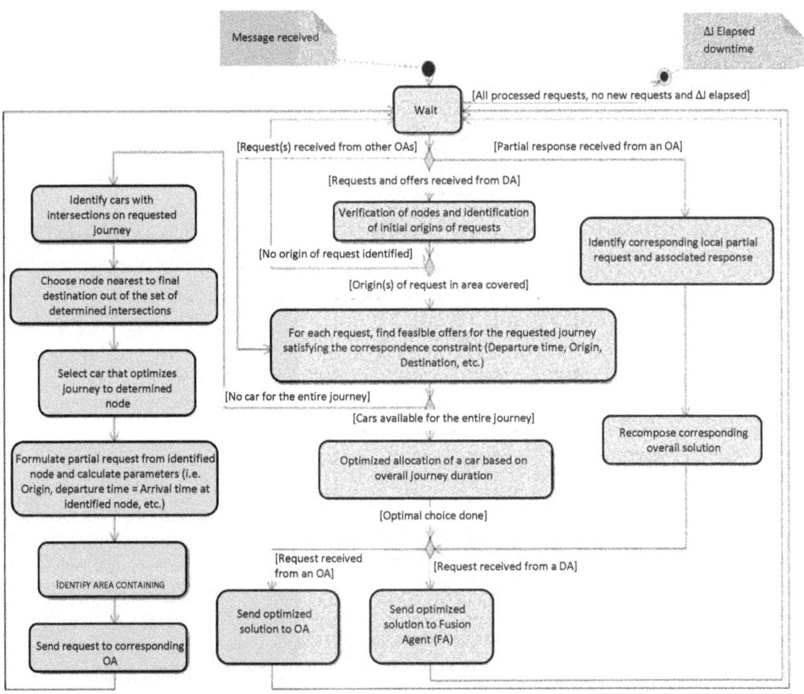

Figure 1.9. *An OA for the optimized distributed processing of requests based on ODAVe*

All operations with the appropriate modeling are detailed throughout algorithms 1.5 and 1.6 below.

Algorithm 1.5. Optimized Distributed Allocation of Vehicle to users (ODAVe)

Data: R_{AO_p} : Requests whose origins are located in zone A_p incumbent upon OA_p; O(t) I(t): Set of offers, TDT : Tolerable Delay Threshold, AIS: Average Itinerary Speed, t : system time, $\{\omega_1, \omega_2\}_R$

Result: S : Set of solutions for R

Initialization: Solution $S \leftarrow \phi$

For All request r such that $r \in R_i$ Do

$S_i \leftarrow OGI(r_i)$

If $S_i = \phi$ Then

$IN \leftarrow IN \bigcup Nearest\_Node(I(t), r_i^-)$

Repeat

$NP \leftarrow Nearest\_Node(IN, r_i^-)$

$IN \leftarrow IN - NP$

$r_{i_1} \leftarrow Establish\_\mathrm{Re}quest(r_i^+, NP, P_i, d_i, a_{NP})$

$S_{i1} \leftarrow OGI(r_{i1})$

$Z_c \leftarrow Area\_Containing(NP, r_i^-, P_i, d_{NP}, a_{r_i -})$

$r_{i_2} \leftarrow Establish\_\mathrm{Re}quest\left(NP, r_i^-, P_i, d_{NP}, a_{r_i-}\right)$

Send r_{i_2} to OA_c $\{OA_c$ is the OA responsiblefor $Z_c\}$

$S_{i2} \leftarrow ODAVe_{AO_c}\left(r_{i2}\right)$

IF $S_{i2} = \phi$ Then

$S_{i2} = \phi$

Otherwise

$S_i = S_{i1} \oplus_{i2}$

End If

Until $S_i \neq \phi$ Or $IN \neq \phi$

End If

$S \leftarrow S \bigcup S_i$

End For

Send S to AF

Algorithm 1.6 below is relative to the function searching for an offer that may respond to the desired path in its entirety.

One option called PS (*Possible Solution*) in our algorithms is thus considered:

$$SP = \left(Departure\_Po\text{int}, Arrival\_Po\text{int}, V_j, Pl_j, Duration, d, a, Fitness\right).$$

Algorithm 1.6. Overall Optimized Itinerary (OOI)

Data: R_i : A specific request included in zone Z_p incumbent upon OA_p; $O(t)$: Set of offers, t : system time, $\{w_1, w_2\}_R$

Result: S_i : Overall optimized solution for R_i

Initialization: Solution $S_i \leftarrow \phi$, $SP \leftarrow \phi$: set of possible solutions

For All offer o_j such that $o_j \in O(t)$ Do

  If $R_i^+ \in IT_{v_j}(t)$ Then

    If $R_i^- \in IT_{v_j}(t)$ Then

      Calculate$(W_t, T_t, Duration, d, a, Fitness)$

        If $(Check\_Constraints(d_i, Pl_j, d, a))$ Then

          $SP \leftarrow SP \bigcup (R_i^+, R_i^-, V_j, Duration, d, a, Fitness)$

        End If

      End If

    End If

  End For

$S_i \leftarrow Choose\ SP\ of\ SP\ that\ optimizes\ Fitness\ Function$

Return S_i

1.6. Implementation and deployment of a dynamic car-pooling service

In this phase, it is necessary to specify how the intervening agents will be deployed in the overall process of our approach. The distribution of the agents is of the utmost importance as it directly affects the cost of communications, monopolization of resources and optimization of resource consumption, etc. It implies specifying the correct way to carry out their deployment in order to optimize the use of resources and save time. The direct impact on the costs in terms of communication stems from the fact that the distribution of agents can promote their optimization or conversely, degrade their quality. Besides the fact that the sending and receiving of a message depends primarily on the type of network in which agents are deployed, it is also necessary to optimize (time) costs of transmission of messages between agents by finding good distribution thereof.

Our approach, as we have defined it, inherits characteristics from the different agents that make it up. In the main optimization algorithms at the heart of the two processes following which we approach the issue of CDOs, we can note that an agent is only created and only intervenes if a need is expressed. As such, we can make our strategy only involve UIAs that relate to users who issued car-pooling requests in our platform. Also, in this context, the principle of decomposition of the way in which it was approached joins with the idea of an "*economy of agents*" by creating OAs only on PZs that were accomplished at the very beginning of the process. An *IZ*, conversely, would only come into play if necessary, and the responsible *OA* is only created if the decomposition of a request requires this (when the intermediate point joining the two partial requests can be found in the IZ concerned). Subsequently, in addition to the benefits of the concept agent, the adopted resolution methodology, whatever the relation to the principle of decomposition or to the distributed optimization methods, brings a significant advantage in optimizing communication costs.

Furthermore, independent of the network topology as well as its nature, the capacity of communication that the agents in our system are equipped with offers the possibility of adaptation and a wide variety of deployment configurations.

In what follows, we present the architecture adopted for our system, and the implementation of the system (ODCCA) which shows different scenarios for the platform execution thus established, called *DOMARTiC* (Distributed Optimized approach based on the Multi-Agent concept to set up a Real-Time Car-pooling service). Moreover, in each test carried out, a representation of the geographical network served was performed thanks to real mapping with integrated software CartoCom.

For the implementation of our ODCCA system or DOMARTiC, the JADE (Java Agent DEvelopment framework) platform [BEL 03; BEL 99] was chosen as the development platform, allowing for more compatibility between agents, total interoperability, and a secure exchange of communication.

1.6.1. *Deployment of ODCCA: choosing a hybrid architecture*

In a centralized architecture, all agents can be found on a central server to which all car-pooling requests and offers are redirected. This configuration has the advantage of considerably reducing the costs incurred by the communication between the agents of our system in the same place and hence avoiding a pointless exchange with a distant network. However, this configuration has the disadvantage of making the agents work sequentially, despite the fact that they are supposed to work in parallel to save time and therefore improve performance. On the other hand, the distributed architecture also has major advantages of which the main objective concerns the optimization of the execution time with the launch of several processes in parallel for the treatment of different processes, including, in our case, the dynamic optimized allocation. These characteristics represent a considerable improvement in the execution time of our system, which allows us to stay in the field of approach dynamics and achieve high performance in real-time systems.

To do this, and thus take advantage of one or the other of the two approaches, we jointly consider both configurations centralized and distributed in a single "*hybrid architecture*". As can be seen in Figure 1.10, the deployment of our request involves several stations to be interconnected through a local network, thus avoiding wasting the time gained by the introduction of parallel processes in long exchanges between agents dispatched on a larger network. In addition to the parallel processing and exchange in local network that supports the optimization of the execution time and thereby increase system performance, we also limit the physical resources used. For example, a group of three machines (Figure 1.10) shows greater performance than if it were bigger because they each host multiple agents, and the exchanges thus done internally positively affect the performance of the request thus deployed.

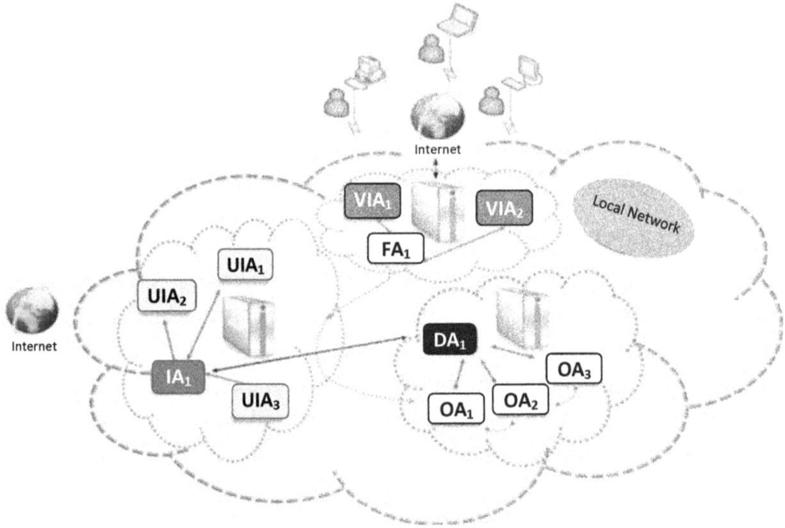

Figure 1.10. *Deployment of our ODCCA system*

In this architecture, an example of deployment of instantiated agents in the context of the completion of ODCCA shows dispatching on three different machines connected through a local network, each with a quad-core processor, a principle and technology that we use to optimize flow while reducing the cost.

Based on this architecture and the development platform, the software architecture is well defined and fulfills a separation between presentation layers, processing and data while setting up different forms of interaction between them. To demonstrate the composition, we present the MVC model that meets the requirements of our system.

1.6.2. *Layered architecture*

In this section, we present the software architecture that represents its composition components or entities involved in the overall operation. We propose to set up our system to adopt the *MVC* architectural model, the basic features of which are explained in the next section.

– MVC model

Our system is modeled as a layered architecture involving three different implementation layers of our system, namely a Model layer, a View layer and Controller layer. More generally called *MVC* (*Model–View–Controller*), it is a design pattern designated rather as an architectural design based on the separation of data (the *Model*), the Human–Machine Interface (the *View: HMI*) and control logic (the *Controller*). The three layers are distinguished as follows.

a) The Model

Specific to the data of a request, this represents the structure but also defines the interactions with the database and the processing thereof.

In our request, we specifically defined the characteristics of each of the agents involved in the implementation of the dynamic optimized allocation method. It should be noted that in the context of *MVC*, we can notice that the *Information Agent* (*IA*) must collect all the updated data required for processing to be performed. Thus it ensures its role to represent all data related to requests and offers of car-share.

b) The View

It covers everything related to interface with the user (*HMI: Human Machine Interface*), meaning that with which it interacts. No processing is done at this level except simply the display of data that it receives from the model layer. Multiple views can be initiated to represent the same data from a single model. In our case, for example, it is essentially views relative to specific interfaces for car-sharees (where the UIA is involved) and those relative to car-sharers (where the VIA is involved).

c) The Controller

This special layer is responsible for managing the interface and interaction between the *Model* and the client. It is therefore responsible for interpreting the request of the user in order to send the view to which it corresponds. The controller performs synchronization between the model and view layers and can thus generate events when the model changes and launch an update of the view already being displayed. In this context, we note that this is the *Fusion Agent* (*FA*) that provides information on all the changes generated on the model layer data (managed by the *IA*). The modifications, made by the *DA* and *OA*, will subsequently be dispatched to different instantiations of the *UIA* and *VIA* with a necessary identification of the corresponding user support so as to express the appropriate interface. In addition, the *DA* and *OA* also play a supervisory role over the views that we cannot ignore. They are actually responsible for the decomposition of the field in the first case and of the generation of optimized itinerary solutions in the second case. Different representations of the modifications generated on the graph that represents the network data thus emerge as a result of the implementation of these various stages of decomposition and search for allocations. A different view of the network can be displayed on a separate interface.

A three-layer architecture, Interface (Model), Application (Controller) and Data (Model), applied to our multi-agent model is shown in Figure 1.11. The latter shows a representation of the layers model with the model Agent proposed in ODDCA. This representation is consistent with the specifications of our problem, the chosen model and the principles and set and achieved treatments.

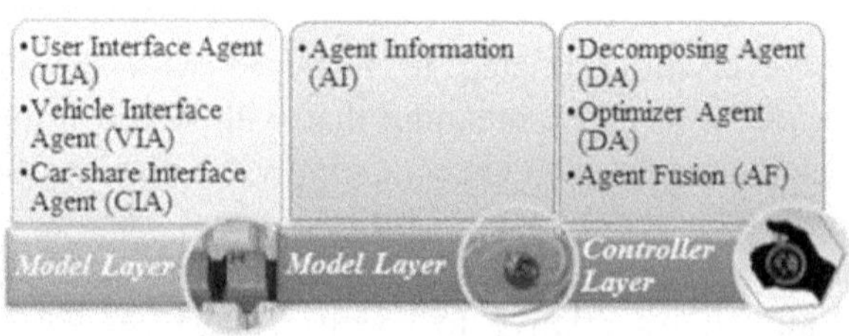

Figure 1.11. *System architecture and involvement of agents in each layer*

Figure 1.12 illustrates the model of the different flows showing exchange opportunities between the different layers.

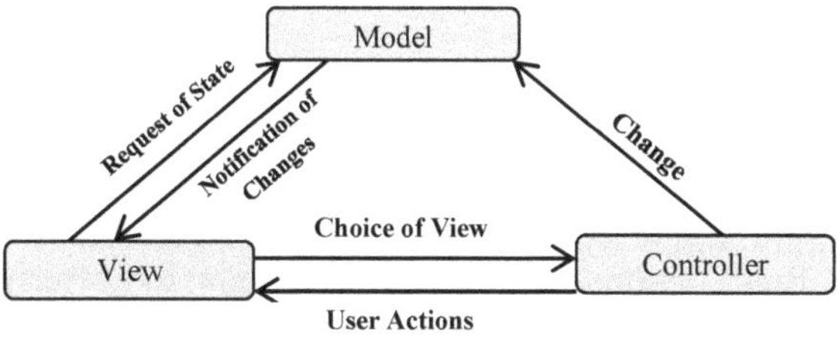

Figure 1.12. *Interactions between the layers of the MVC model*

By implementing our application following this architectural style, we benefit from its main advantages which are:

– the ability to change a layer without affecting the others since they are clearly separated and can be modified without affecting the rest of the layers in addition to the synchronization of the views.

– API JDBC

To access the data system, we made use of the *API* (*Application Programming Interface*) *JDBC* (*Java DataBase Connectivity*) that allows our request to access a database where all the information required for the current processing is stored.

JDBC technology is used to connect specific computer requests to a data server through the manipulation of existing data. It consists of a set of classes allowing the development of these types of request, providing them with the ability to connect to the DataBase Management Server (*DBMS*).

– CartoCom

The very imposing dynamic aspect in the context of total and continuous mobility with respect to the transport sector is very conducive to the allocation ofconflict traps that would no longer be in agreement when issued to car-sharees and car-sharers. Processing such evolving and inconsistent data from one moment to the next is a choice that must be assumed in our system. It is within this framework that we have used the special services of a specific software package called CartoCom by the company Bayo[1].

Equipped with GPS devices, this platform facilitates the acquisition of geographic data and provides continuous tracking of users connected to the system. CartoCom has been ingrained into our platform as a *GIS* unit responsible for recovering data transmitted by users from their origins to their destinations by offering multiple services in parallel, which we took advantage of to achieve our goals. These mainly revolve around the desire to make our system full-scale; the mapping provided by the CartoCom software has enabled us to achieve, in a first instance, the desired objectives from this perspective and in a second instance, integrity and efficiency. Indeed, the services offered by this software are strongly conducive to "*correct*" manipulation free from conflicts, thus helping us to meet the real-time challenge. These services mainly include:

– The continuous geolocation of users of our system, whether they are pedestrians or drivers.

– The geolocation in frequency of scripts also allows full information on the movements of users for the creation of full traceability of the different routes taken by car-pooling or not.

– With CartoCom, our system favors the development of a continuous dialogue with users, taking advantage not only of the role of agents in this regard but of continually having essential information about their movements.

– Another service provided by CartoCom concerns the provision by the ODCCA system of real mapping on which users and their origins and destinations can be seen.

Figure 1.13 displays the solutions generated that represent the matching itineraries via CartoCom.

1 http://www.bayo.com/.

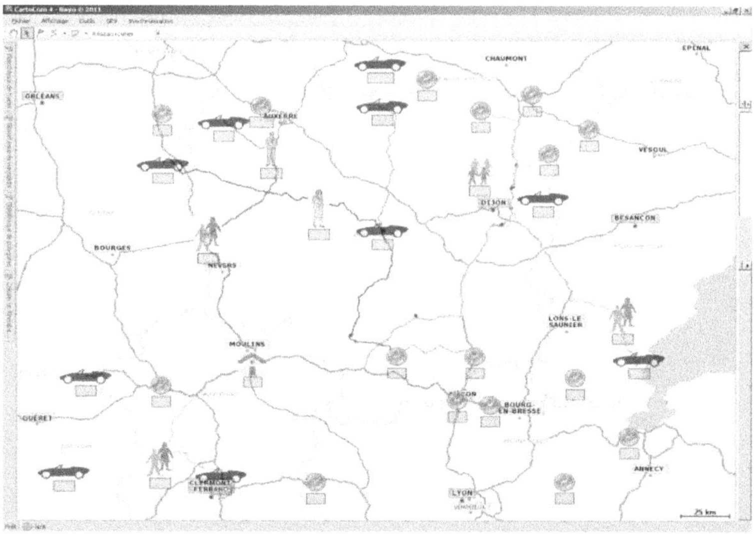

Figure 1.13. *Display of the solution itineraries. For a color version of the figure, see www.iste.co.uk/hammadi/multitransport.zip*

1.6.3. *Testing and implementation scenario*

To demonstrate the validity of the results generated through services provided by the DOMARTiC platform, implementation scenarios will be presented.

1.6.3.1. *User data*

Presented in Table 1.1 is the set of "*pedestrian*" users not yet allocated to a set of pseudo-simultaneous requests received at time *t=9:10*. It is the *IA* that ensured the acquisition. Having been created in the same instant as when the first user request was received, during waiting time $\Delta\varepsilon$ = *3 seconds*, the *IA* gathers information on the requests likely to occur in this time. It is assumed that there will be six and the specifications mainly related to the number of passengers (desired earliest departure and latest arrival time) are also presented.

User identifier	Request Origin	Request Destination	Number of passengers	d	a
U_1	U_1^+	U_1^-	1	9h10	10h40
U_2	U_2^+	U_2^-	2	9h10	10h
U_3	U_3^+	U_3^-	3	9h10	10h15
U_4	U_4^+	U_4^-	4	9h10	10h10
U_5	U_5^-	U_5^-	5	9h15	10h00
U_6	U_6^+	U_6^-	6	9h10	10h30

Table 1.1. *User requests at time t=9:10*

Conversely, Table 1.2 shows the different settings for vehicles emitting car-pooling offers.

Vehicle identifier	Offer Origin	Offer Destination	Number of passengers	Intermediate Destinations	d	a
C_1	C_1^+	C_1^-	1	$(C_5^+, C_8^-, U_3^-, C_7^-, C_5^-)$	-	10h15
C_2	C_2^+	C_2^-	3	$(U_4^+, ID_1, U_6^+, U_4^-)$	-	13h
C_3	C_3^+	C_3^-	4	(U_1^+, U_2^+, ID_1)	-	11h
C_4	C_4^+	C_4^-	3	(U_6^+, ID_1)	-	12h15
C_5	C_5^+	C_5^-	4	(C_8^-, U_3^-, C_7^-)	-	11h15
C_6	C_6^+	C_6^-	8	(U_5^+, U_6^+, U_5^-)	-	12h10
C_7	C_7^+	C_7^-	2	$(C_4^-, ID_1, U_1^-, U_2^-)$	-	12h
C_8	C_8^+	C_8^-	1	(U_1^+, U_3^-, C_5^+)	-	9h45
C_9	C_9^+	C_9^-	3	$(C_4^-, ID_1, U_1^-, U_2^-, C_7^-, U_6^+, U_5^-)$	-	13h15
C_{10}	C_{10}^+	C_{10}^-	2	-	-	13h

Table 1.2. *Car-pooling offers at time t=9:10*

Figure 1.14 shows the layout of nodes on a real map as generated by the CartoCom tool.

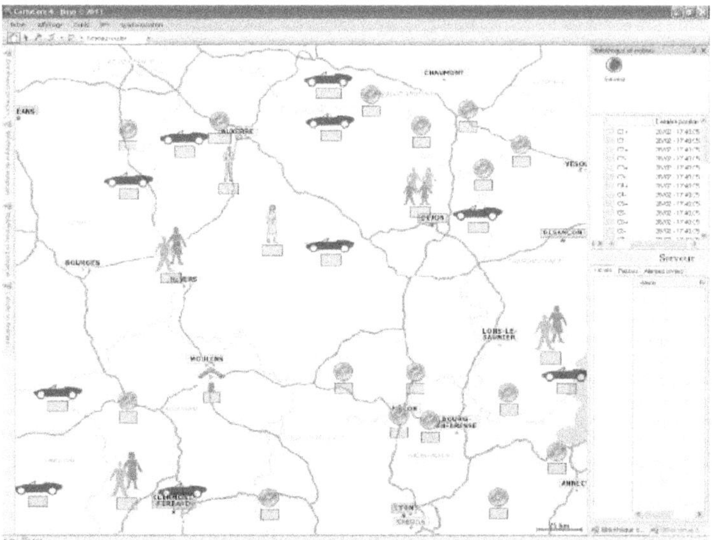

Figure 1.14. *Positions of the origins and destinations of pedestrians and vehicles on a real map. For a color version of the figure, see www.iste.co.uk/hammadi/multitransport.zip*

1.6.3.2. *Processing and allocation*

Processing really begins from the moment the various data are sent by the *IA* to the *DA* to undergo a first classification that highlights areas of data concentration, and where it should start the allocation process favoring those that are less dense and that do not contain request origins.

a) *Requests and primary areas*

The graph shown in Figure 1.15 relates to the modeling of pedestrian information (data in Table 1.1). In this graph, itineraries linking origins to desired destinations were constructed using the CartoCom tool, giving rise to the shortest journeys.

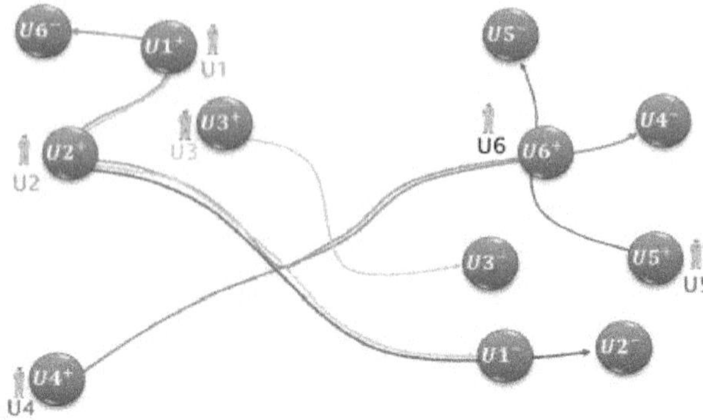

Figure 1.15. *Representation of users (pedestrians) and their itineraries. For a color version of the figure, see www.iste.co.uk/hammadi/multitransport.zip*

The Primary Zones (PZ) are established by the execution of algorithms named *CPZ* (*Creation of Primary Zones*). Firstly, the distances between the places of departure and arrival must be calculated. The distances calculated thus are used to determine membership or not of a set of nodes to the same area, compared to the determined diameter and for calculating the position of the centers of areas.

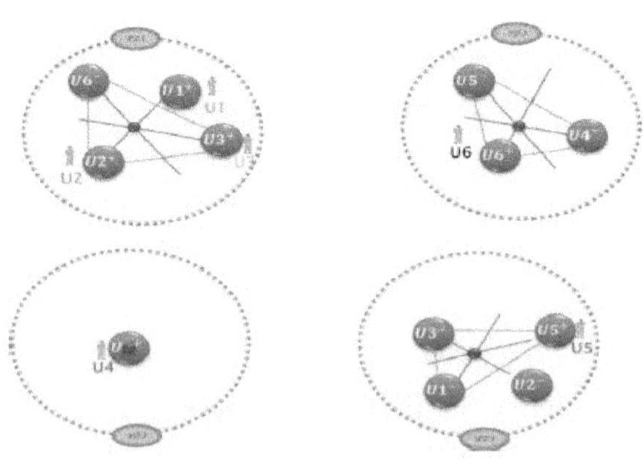

Figure 1.16. *Primary Zones*

b) ***Cars and intermediate zones***

As was done for car-pooling requests, almost the same procedure is followed in this second stage of decomposition.

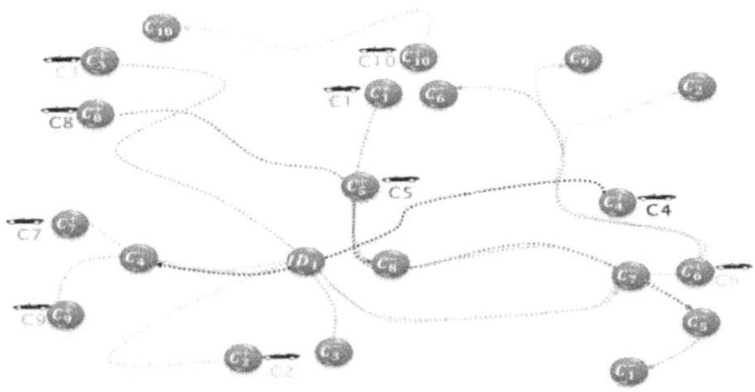

Figure 1.17. *Graph of origins and destinations of cars. For a color version of the figure, see www.iste.co.uk/hammadi/multitransport.zip*

Subsequently, a graph superposition of the two planes (pedestrian plane and car plane) in space is made creating a fusion of the two graphs to bring out the similarities and especially the correspondence between journeys.

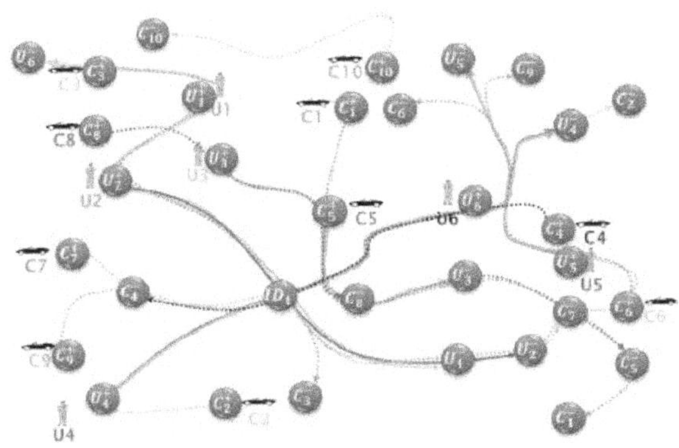

Figure 1.18. *Superposition of car-sharers and car-sharees. For a color version of the figure, see www.iste.co.uk/hammadi/multitransport.zip*

On the basis of this modeling and the Primary Zones created by the first phase of decomposition of the terrain (based on user requests only), the Intermediate Zones *IZ* will be gradually developed according to the *CIZ* (*Creation of Intermediate Zones*) algorithms previously developed and established. These first seek possible allocations of newly constructed nodes to previously developed *PZs*, and create new areas for those not allocated to the same neighborhood principle inspired by data classification methods (Figure 1.19).

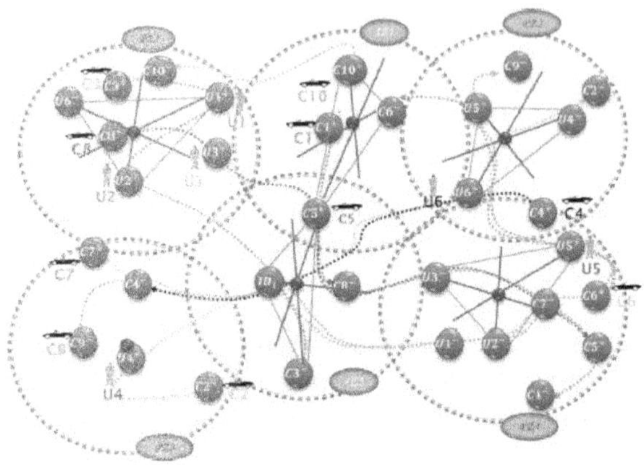

Figure 1.19. *Intermediate Zones and network coverage. For a color version of the figure, see www.iste.co.uk/hammadi/multitransport.zip*

1.6.3.3. *Some calculations*

Indeed, four *OAs* are created and launched at the same time to execute each *ODAVe* on requests whose origins lie in the geographical coverage area (PZ). Some examples, illustrative of calculations regarding intervening agents, are detailed in Figures 1.20–1.23 with the necessary data, particularly in relation to the distances between the nodes as well as the weights of the Fitness function specific to users affected by the request undergoing processing in each example.

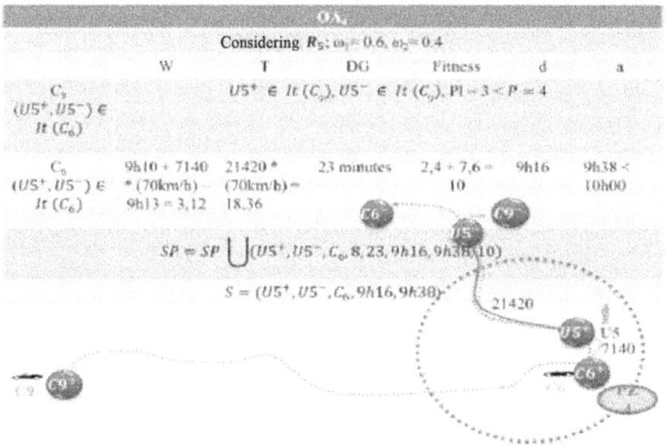

Figure 1.20. *Processing of R5. For a color version of the figure, see www.iste.co.uk/hammadi/multitransport.zip*

The R_5 example has two choices and coincidentally checks correspondence constraints. However, one of the two choices does not meet the feasibility of the offer.

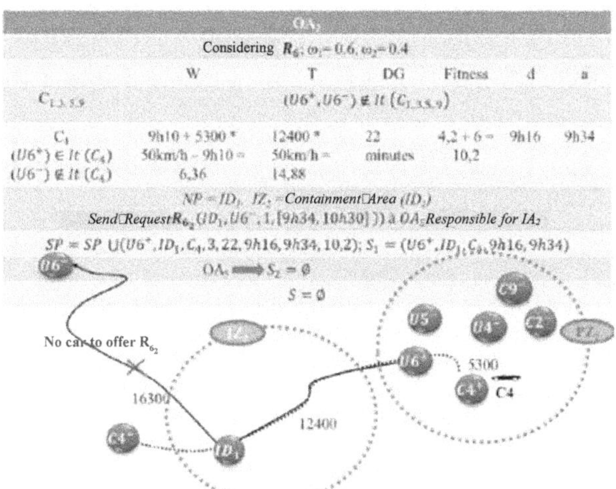

Figure 1.21. *Processing of R6*

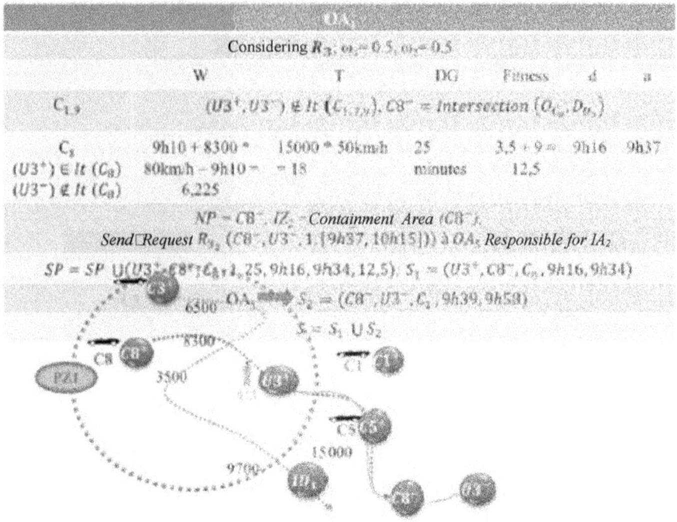

Figure 1.22. *Processing of R3. For a color version of the figure, see www.iste.co.uk/hammadi/multitransport.zip*

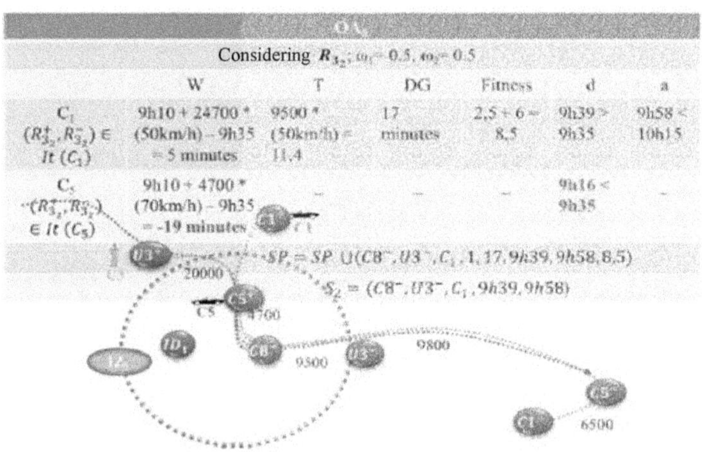

Figure 1.23. *Processing of the second part of R3. For a color version of the figure, see www.iste.co.uk/hammadi/multitransport.zip*

The last three figures highlight the collaborative operation of the agents for the treatment of R6 and R3 which involves a coalition between the OA2 and OA5 agents for the first part and OA1 and OA5 for the second part.

1.6.3.4. *Displaying the results*

Figure 1.24 displays the generated solutions that represent the corresponding itineraries via CartoCom.

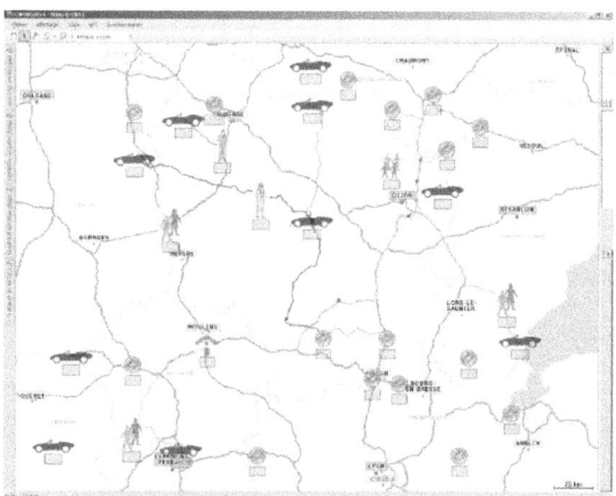

Figure 1.24. *Display of the solution itineraries. For a color version of the figure, see www.iste.co.uk/hammadi/multitransport.zip*

1.6.3.5. *DOMARTiC: the interface: execution example*

In what follows, we present different interfaces (Figures 1.26–1.27) generated by directly testing our application (DOMARTiC). A random generation module of "*mobiles*" (pedestrians and cars) has been developed to ease testing, all the while handling real geographic data. In Figure 1.27, we can see the different exchanges between the agents involved, especially in the context of a coalition.

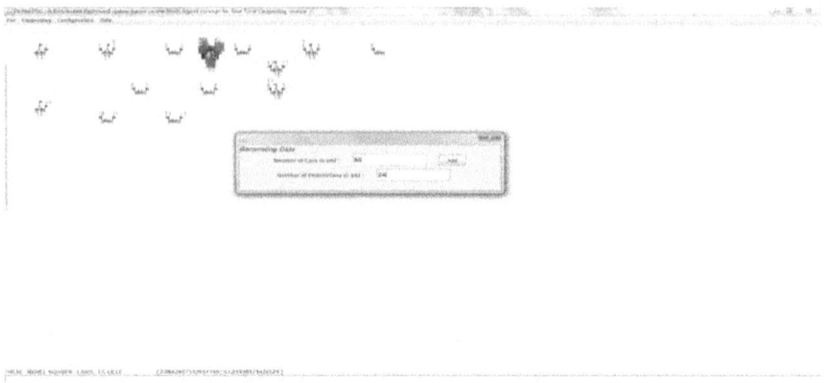

Figure 1.25. *Random data generation*

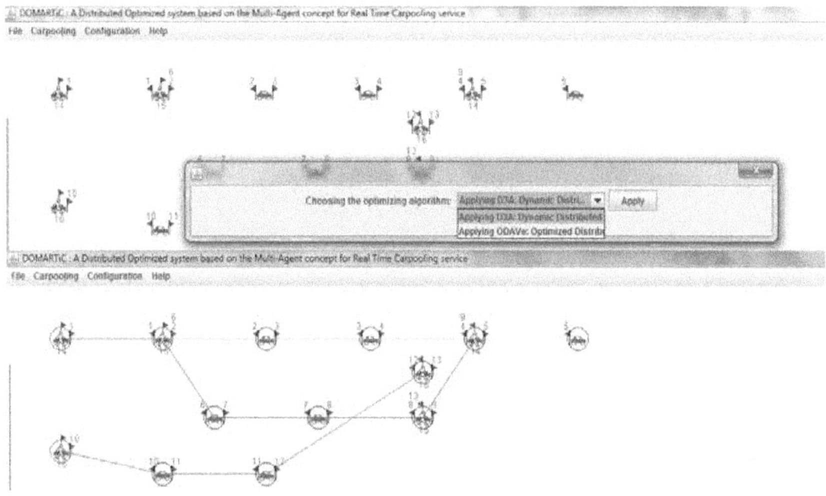

Figure 1.26. *Example of an execution. For a color version of the figure, see www.iste.co.uk/hammadi/multitransport.zip*

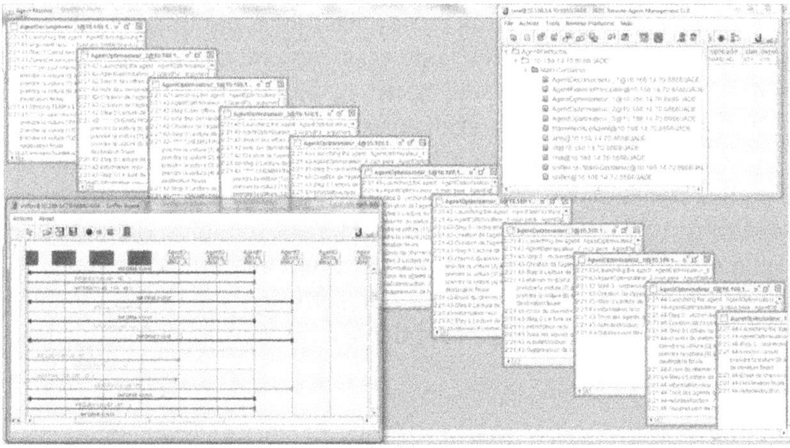

Figure 1.27. *Coalition between OA and recomposition of itineraries*

1.7. Conclusion

In this chapter, we presented the implementation of a dynamic distributed car-pooling system, ODCCA, for which we demonstrated the exponential complexity. We focused our efforts on the development of an effective resolution strategy leveraging a mixture of concepts, namely, multi-agent systems and optimization. A combination of these two founding concepts was the main contribution of this work since it exploits the benefits of both in a perfect and harmonious combination allowing a distributed artificial intelligence context to rule. Two approaches have been proposed in our system (ODCCA), the first proposal implements a Dijkstra distributed dynamic allocation (D3A) algorithm which was conducted to address the problems outlined previously and provide a complex service of polynomial order. Still in the context of multi-objective optimization, the second aspect concerns a heuristic combining even more optimization criteria. ODAVe implements a distributed optimization with the same goals as the first approach (i.e. to reduce complexity) by taking into account the lexicographical order on the set of optimization criteria (comfort and overall travel time).

To consolidate these choices in pursuit of a performance that will make ODCCA a competitive large-scale system, appropriate technical choices were required. To be noted in particular in this context: a JADE platform facilitating exchanges, a deployment model involving a hybrid architecture juggling between a centralized architecture and the distributed local network, an *MVC* model preventing any possibility of change or integration of new modules (i.e. statistics, planning and anticipation of travel, etc.) and a *GIS* module integrated via an interface with the CartoCom software. All these combined techniques and technologies lead the system, developed on this basis, towards efficiency.

1.8. Bibliography

[ALA 06] ALAMI J., EL IMRANI A., "Algorithme culturel parallèle pour l'optimisation multimodale", *Journal électronique d'intelligence artificielle*, June 2006.

[ALT 04] ALTMAN E., WYNTER L., "Equilibrium, games and pricing in transportation and telecommunication networks", *Networks and Spatial Economics: special issue on crossovers between transportationand telecommunication modelling*, 7-21, 2004.

[AMM 07] AMMOR O., RAISS N., SLAOUI K., "Détermination du nombre optimal de classes présentant un fort degré de chevauchement", *La revue MODULAD*, 37, 31-42, December 2007.

[ANI 92] ANILY S., HASSIN R., "The swapping problem", *Networks*, 22, 419-433, 1992.

[ATA 10] ATAHRAN A., LENTE C., T'KINDT V., "Transport à la demande et 'dynamic dial and ride problem': liens et état de l'art", *ROADEF 2010, 11e Congrès annuel de la société française de Recherche Opérationnelle et d'Aide à la Décision*, 2010.

[ASH 98] ASHOK K., Estimation and prediction of time dependent origin-destination flows, PhD Thesis, MIT, 1998.

[BAR 98] BARNHART C., JOHNSON E.L., NEMHAUSER G.L., SAVELSBERGH M.W.P., VANCE P.H., "Branch-and-price: 'column generation for solving huge integer programs'", *Operations Research*, 46, 316-329, 1998.

[BAR 07] BARNHART C., LAPORTE G. (eds), *Transportation*, Elsevier, Amsterdam, 2007.

[BEL 99] BELLIFEMINE F., POGGI A., RIMASSA G., "JADE – A FIPA-compliant agent framework", *Proceedings of the The Fourth International Conference and Exhibition on the Practical Application of Intelligent Agents and Multi-Agents*, 1999.

[BEL 03] BELLIFEMINE F., CAIRE G., POGGI A., RIMA G., *JADE White Paper*, 3(3), 2003.

[BEN] BENOÎT R., *Communauto*, http: //www.communauto.com/.

[BER 10] BERBEGLIA G., CORDEAU J-F., LAPORTE G., "Dynamic pickup and delivery problems", *European Journal of Operational Research*, 202(1), 8-15, 2010.

[BOR 97] BORNDORFER R., GROTSCHEL M., KLOSTERMEINER F., KUTTNER C., "Telebus Berlin: vehicle routing scheduling in a dial a ride system", *Konrad Zuse Zentrum fur Information Technik Berlin*, 1997.

[BOU 02] BOULOGNE T., ALTMAN E., POURTALLIER O., "Mixed equilibrium in multiclass routing games", *Annals of Operation Research*, 109, 279-291, 2002.

[BRO 03] BROTCORNE L., LAPORTE G., SEMET F., "Ambulance location and relocation problem", *EJOR 147*, 451-463, 2003.

[CED 01] CEDER A., "Efficient timetabling and vehicle scheduling in public transport", in VOSS S. and DADUNA J.R. (eds), *Computer-Aided Scheduling of Public Transport*, 37-52, 2001.

[COR 02] CORDEAU J.-F., GENDREAU M., LAPORTE G., POTVIN J.Y., SEMET F., "A guide to vehicle routing heuristics", *Jour. Op. Res. Soc.*, 53-5, 512-522, 2002.

[COR 03a] CORDEAU J.-F., LAPORTE G., "A tabu search heuristics for the static multi-vehicle dial a ride problem", *Transportation Research B*, 37, 579-594, 2003.

[COR 03b] CORDEAU J.-F., LAPORTE G.,"The dial a ride problem: variants, models and algorithms", *40R1-2*, 89-101, 2003.

[COR 07] CORDEAU J.-F., LAPORTE G., "The dial-a-ride problem: models and algorithms", *Annals of Operations Research*, 153, 29-46, 2007.

[DUM 91] DUMAS Y., DESROSIERS J., SOUMIS F.,"The pickup and delivery problem with time windows", *European Journal of Operational Research*, 54(1), 7-22, Elsevier, September 1991.

[FEK 10] FEKI M.F., Optimisation distribuée pour la recherche des itinéraires multi-opérateurs dans un réseau de transport co-modal, PhD Thesis, Ecole Centrale de Lille, 2010.

[FLE 09] FLEURY E., *Réseaux de Capteur*, Dunod, 2009.

[GIL 02] GILLEN D., HAYNES M., Public and private benefits in intelligent transportation systems/commercial vehicle operations: electronic clearance and supply chain management, California Partners for Advanced Transit and Highways (PATH), Research Reports: Paper UCB-ITS-PRR-2002-18, 2002.

[HEU 08] HEUVEL A.P.R., AKKER J.M. and KOOTEN NIEKERK M.E., Integrating timetabling and vehicle scheduling in public bus transportation, Technical Report, Department of Information and Computing Sciences, Utrecht University, The Netherlands, February 2008.

[HIG 00] HIGGINS L., LAUGHLIN J.B., TURNBULL K., "Automatic vehicle location and advanced paratransit scheduling at Houston METROLift", *Proc Transport. 2000 Research Board Conference*, 2000.

[HIZ 08] HIZEM M.M. Recherche de chemins dans un graphe à pondération dynamique: application à l'optimisation d'itinéraires dans les réseaux routiers, Thesis, Ecole Centrale de Lille, 2008.

[HU 05] HU J., SHI X. ,SONG J., XU Y., "Optimal design for urban mass transit network based on evolutionary algorithm", *Lecture Notes in Computer Science 3611*, 1089-1100, 2005.

[IGO] IGO, http ://www.igocars.org/how.

[IOR 07] IORI M., SALAZAR GONZÁLEZ J.J., VIGO D.,"An exact approach for the vehicle routing problem with two-dimensional loading constraints", *Transportation Science*, 41(2), C.S. Journals, 253-264, May 2007.

[JAG 03] JÄGER W., KREBS H.-J. (eds), *Mathematics Key Technology for the Future – Joint Projects between Universities and Industry*, Springer, 2003.

[JES 08] JESPERSEN-GROTH J., POTTHOFF D., CLAUSEN J., HUISMAN D., KROON L., MAROTI G., and NIELSEN M.N., "Disruption management in passenger railway transportation", *Computers & Operations Research'*, 2008.

[KAM 07] KAMOUN M.A., Conception d'un système d'information pour l'aide au déplacement multimodal: Une approche multi-agents pour la recherche et la composition des itinéraires en ligne, PhD Thesis, Ecole Centrale de Lille, 2007.

[KON 06] KONINGS R., PRIEMUS H., NIJKAMP P., (eds.), *The Future of Automated Freight Transport – Concepts, Design and Implementation*, Edward Elgar Publishing, 2006.

[KOH 07] KOHL N., LARSON A., LARSEN J., ROSS A., and TIOURINE S., "Airline disruption management – Perspectives, experiences and outlook", *Journal of Air Transport Management*, 13(3), 149-162, 2007.

[LEC 06] LECCHINI A., GLOVER W., LYGEROS J., MACIEJOWSKI J., "Monte Carlo optimisation for conflict resolution in air traffic control", *IEEE Transactions on Intelligent Transportation Systems*, 7(4), 470-482, 2006.

[LIL] LILAS AUTOPARTAGE, http ://www.lilas-autopartage.com/savoir.html.

[LIS] LISELEC, http://www.energie-cites.org/db/la_rochelle_569_fr.pdf.

[MCB 83] MCBRIDE M.E., "Spatial competition and vertical integration: Cement and concrete revisited", *The American Economic Review*, 73(5), 1011-1022, 1983.

[MOB] MOBILITY, http ://www.mobility.ch/pages/index.cfm?srv =cms&pg =&dom=6&prub=526&rub=527.

[POC 06] POCHET Y., WOLSEY L., *Production Planning by Mixed Integer Programming*, Springer, 2006.

[PSA 95] PSAFARATIS H.N., "Dynamic vehicle routing: status and prospects", *Annals of Operations Research 61*, 143-164, 1995.

[SAV 95] SAVELSBERGH M.W.P., SOL M., "The general pickup and delivery problem", *Transportation Science*, 29, Cite Seerx, 17-29,(1995).

[SCH 78] SCHULER R.E., HOLAHAN W.L., "Competition vs. vertical integration of transportation and production in a spatial economy", *Papers in Regional Science*, 41(1), 209-225, 1978.

[STO 00] STONE J.R., AHMED T., NAVALENKO A.M. "Internet based decision support for advanced public transportation systems", *Proc Transport, Research Board Conf.*, 2000.

[TAM 91] TAMIR A., "On the core of network synthesis games", *Math Programming 50*, 123-135, 1991.

[TER] TERESE, CERTU-Lyon, Internal Communications

[VAN 04] VAN DAM K.H., OTTJES J.A., LODEWIJKS G., LUKSZO Z.V., WAGENAAR R.W., "Intelligent infrastructures: distributed intelligence in transport system control - an illustrative example", *IEEE International Conference on Systems, Man and Cybernetics*, 5, 4650 – 4654,2004.

[WAR 52] WARDROP J., "Some theoretical aspects of road traffic research", *Proc. of the Institute of Civil Engineering*, II, 1, 325-378, 1952.

[WCC] THE WORLD CARSHARE CONSORTIUM, http://www.cities. worldcarshare.com.

[WU 09] WU J., TAN Y.,"A particle swarm optimization algorithm for grain logistics vehicle routing problem", *ISECS International Colloquium on Computing, Communication, Control, and Management*, 364-367, 2009.

[XIA 08] XIANG Z., CHU C., CHEN H.,"The study of a dynamic dial-a-ride problem under timedependent and stochastic environments", *European Journal of Operational Research*, 185(2), 534-551, Elsevier, 2008.

[ZIP] ZIPCAR, http ://www.zipcar.com/about.

Chapter 2

Simulation of Urban Transport Systems

2.1. Introduction

The transport of people and goods is a fundamental element of modern economies. Given the multitude of players, and their often conflicting goals, it is a complex system or even a "system of systems" [LUZ 10]. It is particularly scarred by the permanent contradiction between the ever-increasing mobility requirements of citizens who tolerate less and less the constraints associated with transportation, and economic and social challenges, including those related to sustainable development.

Unfortunately, the mismatch between transport systems and the demand for transport and transport planning in urban, suburban and regional territories causes many negative consequences: lost time, deterioration of quality of life, environmental nuisance, security risks and ultimately economic and societal losses.

In this chapter, we are particularly interested in transport systems of people in urban spaces, both guided and non-guided, and the simulation of associated processes.

Chapter written by Christian TAHON, Thérèse BONTE and Alain GIBAUD.

In the first part of the chapter, we recall the whole context of simulation of urban transport systems and the different types of models used, particularly microsimulation, that are better suited to the consideration of procedures, constraints and specificities of the urban environment and its interactions with transport systems.

We then give a brief overview of the areas covered by urban simulation, of the different models used and an overview of the software tools used, especially those available on the market.

In the following two sections, we present two simulators developed through research. The first is a simulator of multimodal guided urban transport systems, which was used to validate a prototype system to help control traffic for a bimodal tram–bus network (Transvilles network, see the second part of this chapter). The second is based on the principle of self-organization aiming to improve traffic flow of individual vehicles by exploiting a navigational help system based on early traffic detection.

2.2. Context

The demand for urban transport has changed considerably over the last 20 years and has become more complex. At the same time, urban transport has become a political and strategic issue for cities because of the structural challenges involved in their futures. It appears that these challenges converge towards the goal of sustainable development of cities and they question the integration of specific policies (car parks, roads and urban transport) in urban policies as well as consistency of public choices on urbanization and travel. Beyond these challenges, the possibility of different local players (civil society, transport operators and institutions) influences local public policies [RIE 03].

The importance of urban transport has, in France, led to the set-up of a regulatory framework. Urban Transport Plans (*Plans de Déplacements Urbains* – PDU) were created by the Law of Orientation on Inland Transport (*Loi d'Orientation sur les Transports Intérieurs* – LOTI) of December 30 1982 to streamline the organization of transport, traffic and parking. Thereafter, the Law on Air and the Rational Use of Energy (*Loi sur l'Air et l'Utilisation Rationnelle de l'Energie* – LAURE) of December 30, 1996 amended the plans to create the tools to fight against air pollution. The Solidarity and Urban Renewal law (*Loi Solidarité et Renouvellement Urbain* – SRU) of December 13, 2000 again amended the PDUs in order to articulate them more strongly with urbanization documents. The Urban Transport Plan:

– defines the organization principles of transport of people and goods, as well as those of traffic and parking, within the perimeter of urban transport;

– aims to ensure a sustainable balance between the mobility requirements and ease of access, on the one hand, and environmental and health protection on the other;

– must provide coordinated use of all modes of transport by promoting those which are least polluting and most energy-efficient;

– specifies the design and operational measures to be implemented;

– seeks to moderate the use of the car.

In 2008, the French Ministry of Ecology, Energy, Sustainable Development and the Sea[1] launched a call for "Urban Transport" projects. Presented as a practical implementation of the Grenelle Environment Commitments to favor the development of public transport, it aims to consistently accelerate the development of urban and suburban transport and in particular allow the transport sector to contribute to reducing greenhouse gas emissions.

Moreover, the concept of Intelligent Transport System (ITS) introduced a systemic dimension into the design, planning and control of urban transport systems.

The "Intelligent Transport System" program was launched as an initiative by the US Department of Transport. Its objective is to integrate information and communication technologies with vehicles

1 http://www.developpement-durable.gouv.fr.

and transport infrastructure. It aims for a comprehensive and optimized management of vehicles, transport loads and lanes to improve safety, and reduce the use of personal vehicles, transport times and energy consumption and associated costs [TAY 01].

Many features fall within the scope of the ITS:

– traffic management, which includes sensors, telecommunications, and various operation support systems;

– public transport, including tracking systems and related services, user information services, and on-demand transport services;

– driving aids, such as guidance systems in vehicles, or systems with automatic recognition of traffic signs;

– information for drivers;

– electronic payment systems, with ticketing, electronic toll collection, payment of parking, etc;

– devices intended to promote better compliance of regulations (particularly automatic radar systems);

– systems conceived for road safety, whether as ground equipment (e.g. automatic incident detection), or embedded systems (intelligent speed limiter, automatic emergency call, etc.);

– fleet and freight management, whether the itinerary is done by road, rail, river, etc.

The ITS challenges are diverse, industrial and commercial: improving transport efficiency and safety, particularly road safety, sustainable development, pollution reduction, mobility management, service development. In France, in order to promote a harmonious development of these systems, the ACTIF (*Aide à la Conception de systèmes de Transport Interopérables en France* – Aid for the Conception of Interoperable Transport systems in France) project was implemented [ACT 03].

It is therefore essential to have models that can take into account and simulate all functions induced by the ITS: dynamic traffic management and regulation of flows and networks: intermodality, interactions between systems, information exchange processing, involvement of new technologies (e.g. RFID, etc.) and consequences for the operation, organization, management and integration of new services in the transport systems (on-demand transport, car-sharing, car-pooling, etc.), in particular the issues raised by interoperability between organizations, players and between transport systems.

Modeling and simulation of urban transport systems is the subject of much research and development carried out by many international players (USA, Japan, Europe, etc.) from different backgrounds. In France, for example, the players in the field are, non-exhaustively, university laboratories and engineering schools, ministries, industrial partners (RATP, SNCF, etc.), operators and public transport organizing authorities, INRETS, LET, CNRS, ADEME, etc.

2.3. Simulation of urban transport systems

Urban transport systems are complex systems whose performances are the result of decisions made at different levels of responsibility and by various institutions whose objectives can be contradictory. The design, evaluation and management of such systems requires the use of appropriate models representing dynamic processes and interactions between urban development and transport systems, contributing to the evaluation of alternative strategies [LAB 08; ROU 03].

Figure 2.1 shows an overall modeling process [WAD 04] which comprises a goal-setting phase translated in terms of objectives, determination of strategies and corresponding scenarios, the construction of models to analyze the effects of these scenarios by relying on indicators, and the evaluation of the effectiveness of the policy in achieving these objectives.

The process is naturally iterative insofar as stakeholders will not necessarily agree on the relative importance of each objective, and there are many possible alternatives. The procedure should lead to an agreement on a set of goals and the best political strategy to achieve them. There are multiple purposes for modeling and simulating transport systems, especially in urban areas: planning, management, control, decision, support and strategy implementation [LEB 05].

The simulation of urban transport systems concerns the transport of persons (public transport), goods, waste and hazardous materials. Appropriate models must be developed in conjunction with the simulation of traffic. In this chapter, we consider two types of transport where simulation has become essential: the design and regulation of guided (trams, guided buses, etc.) and non-guided (cars, buses, etc.) people transport systems.

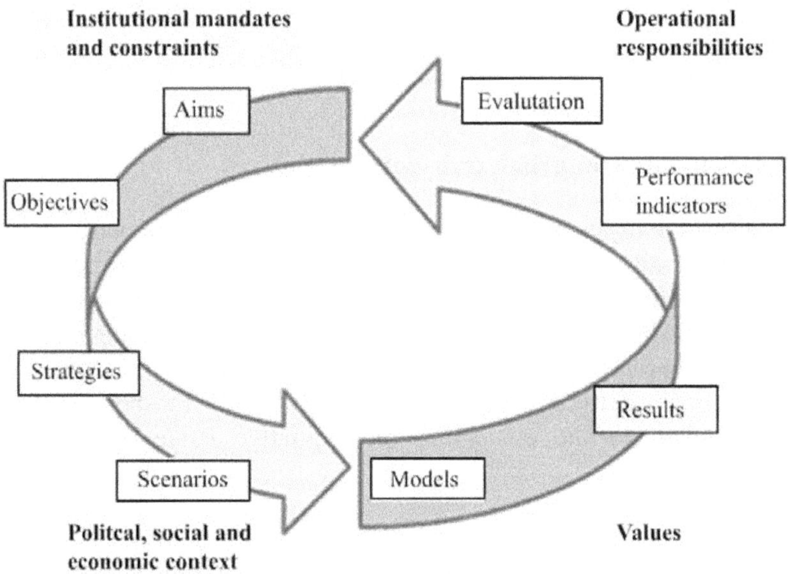

Figure 2.1. *Modeling and urban simulation overall procedure (adapted from [WAD 04])*

2.3.1. *Non-guided transport systems*

The simulation is used for the design of traffic infrastructure, but the main problem to be solved is the increasing congestion of traffic that represents a critical socio-economic problem. Congestion typically leads to an inexorable slowing down of traffic and causes delays that have a significant impact on the quality of life of users and on increased energy consumption. Finally, from an environmental perspective, congestion has an adverse and negative impact due to the increase in pollution that causes a significant increase in economic and environmental costs [GER 08].

One solution is of course to build more roads, but this is relatively limited due to prohibitive costs and lack of available space. It is therefore necessary to implement measures so that the capacity of the existing infrastructure is optimized. In this context, traffic control plays an important role in solving congestion problems and optimizing infrastructure performance. It is currently the only effective way to eliminate, or at least reduce, the effects of congestion, provide effective and sustainable solutions for safe transport and allow a reduction of pollution. Traffic control involves the implementation of a series of actions and measures that can be used individually. These include, for example, access control (urban tolls) by dynamic speed limits, dynamic routing, self-regulation, etc. The objective is not only to reduce congestion but also to respect a number of criteria and constraints such as minimizing the travel time of users. An example of a self-regulation system is the subject of the third part of this chapter.

2.3.2. *Guided transport systems*

The overall problem can be expressed by three classes of specific problems:

– network design;

– their optimization and control, which must ensure intermodality control (network and traffic regulation and taking into account uncertainties and potential disruptions);

– logistics of transport means, which must enable the optimization of performance of available resources.

Each class of problem provides an additional perspective to the others, for the effective resolution of concrete cases.

2.3.2.1. *The design (or redesign) of networks*

Network design must define, on the one hand, paths followed by the means of transport, and on the other hand, the interaction points, or interchanges, between specific networks. The design thus poses two types of problem:

– Technical problems: the technical aspect concerns the definition of the architecture of interchanges. In this context, the objectives are multiple:

- design an architecture providing a compromise between the quality of services rendered to users (including, for example facilitating, reducing and securing paths of products and/or persons between the different means of transport at the interchanges, making the wait more pleasant and shorter);

- ensure maximum operating quality including minimizing waiting time and preserving a maximum of equipment and costs (cost and profitability of the proposed solutions);

- integration into the existing environment which results in planning constraints (integration with existing buildings, consideration of topographical features, etc.).

– Organization problems: these are divided into three specific points:

- the location and design of interchanges, based on the consideration of three types of constraint:

– technical: choosing a location for the best possible fluidity of different flows relative to the various networks and corresponding to a strategic decision point for regulation, avoid congestion, cover the widest possible area, etc.,

– economic: coverage of area reaching the most potential users, optimization of interchange positioning to best limit their numbers while maximizing the coverage area to achieve maximum profitability, etc.,

– social: opening of disadvantaged areas, etc.;

- the size of interchanges: this aims to define static parameters (static in the sense that their development will be considered in the long term and will induce significant structural changes) that will define the maximum capacity that the interchange is capable of supporting, particularly in terms of flow.

This design will cover the different elements of the interchange (internal organization, parking areas for vehicles, queues for passengers, buses, trams, etc.).

2.3.2.2. *Optimization and control*

The use of different network parameters depends, on one hand, on the network characteristics and on the other hand, in the context of intermodality, constraints on other networks according to a mutual influence principle. Similarly, the management parameters of interchanges depend on the operating parameters of different networks. We therefore see a considerable challenge induced by intermodality, optimization and control of interchanges since the decisions not only have a local influence, but also affect the entire network.

In all cases, decisions will be subjected to constraints corresponding to three complementary and inseparable perspectives:

– The operating company: economic profitability objectives, ease of operation, taking into account the technical characteristics of the rolling stock, etc.

– The user in the case of passenger transport (urban networks): quality of service objectives especially in terms of waiting times, security, comfort, itinerary, etc.

– The environment: pollution induced on other networks in the interchange, but also on the wider environment, for example in terms of noise, etc.

It is essential to optimize different transport means (assignment, planning, etc.), but also to cope with disruptions caused by unforeseen events in order to avoid the propagation of uncertainties between networks from a given network. Synchronization is achieved by defining and dynamically adapting the operating parameters of the various networks linked to an interchange, such as frequency of service, waiting time, velocity profiles. The definition of operating parameters must be supplemented by management of rules allowing for dynamic adaptation.

2.3.2.3. *Logistics management of transport means*

The optimization of the operation of the intermodal network and its management also imply that the transport means are available and able to perform their duties with maximum safety and security. In this regard, it implies maximizing the availability of various pieces of equipment involved in the operation of the network:

– static equipment: network infrastructure, maintenance depots and workshops, etc;

– mobile equipment: trains, trams, buses, etc.

Moreover, this optimization can only be achieved if we place ourselves in the context of the lifecycle of the equipment, to consider the technical and economic aspects. This is a logistics support process that must be adopted to ensure a particular organization and management of all support resources (maintenance workshops and repair of mobile equipment, response teams, emergency equipment, etc.).

2.4. The types of modeling

Several types of model are used to simulate urban transport systems and they can be categorized according to various classifications including the nature of the models and the level of detail of the simulation. We briefly present a few.

2.4.1. *Nature of the models*

2.4.1.1. *Continuous – discrete*

Simulation models can be classified according to the nature of the system to simulate (discrete, continuous). We recall that a discrete change in system state occurs at given instants characterized by the appearance of one (or more) event(s) such as arrival of a bus at a station. Between two consecutive events, the state of the system remains unchanged. If the number of events is finished, the simulation is denoted by discrete simulation. If the system state continuously changes with time, described for example by differential equations, the system is continuous and the simulation is said to be continuous.

2.4.1.2. *Deterministic – stochastic*

Another classification concerns the type of process shown: deterministic or stochastic. Deterministic models do not include random variables; interactions between modeled entities are defined by exact relationships (mathematical, logical, etc.). Stochastic models include probability functions. For example, the reaction time of a driver following a vehicle can be modeled by a constant value (deterministic model) or by a random variable (stochastic).

2.4.2. *Macrosimulation, mesoscopic simulation, micro simulation*

Simulation models can also be classified according to the level of detail in which the system is represented. Depending on the purpose of the simulation, modeling levels range from macroscopic (simulation based on the traffic flow, density and speed) to microscopic, which describes the motion of each vehicle.

2.4.2.1. *Macroscopic simulation models*

Macroscopic simulation models consider a group of vehicles and do not consider the vehicles individually. The general flow models, also applicable to fluids, can be used for this type of simulation to determine the traffic state at a given time. This level of aggregation is usually used for static planning models and applied to large geographic areas. Due to their static nature, these models cannot consider short-term changes in traffic conditions and their use is limited.

Most macroscopic models were developed in the late 1960s and early 1970s. TRANSYT, FREQ, FREFLO and META SIMAUT are among the most well-known models [BOX 00].

2.4.2.2. *Mesoscopic simulation models*

Mesoscopic simulation models correspond to a finer level of detail than previous models.

These models can only take into account minor changes in traffic patterns over a short period of time.

Some examples of commercial tools are: Dynamit, Dynemo, Metropolis. More recently, a simulator has been developed for the design of interchanges [KAA 07].

2.4.2.3. *Microscopic simulation models*

Microscopic simulation models take into account the characteristics of each vehicle individually and the interactions between each of them. Given the level of detail required for the microscopic simulation, simulation generally concerns a small geographic area.

The microscopic models are best suited to the simulation of urban transport systems. They can reproduce with sufficient accuracy the different situations of urban traffic in various circumstances, and meet the requirements involved in the consideration of the application of intelligent transport systems [BAR 04; JAY 03]. The most commonly used tools are AIMSUN2[2], PARAMICS[3], HUTSIM[4], DRACULA[5] and VISSIM[6]. Among the tools available for modeling traffic in urban streets, 12 of them were deemed capable of modeling the functions associated with Intelligent Transportation Systems [MIL 03; BER 02; ADA 00].

---

2 http://www.aimsun.com.

3 http://www.paramics-online.com.

4 http://www.tkk.fi/Units/Transportation/HUTSIM.

5 http://www.its.leeds.ac.uk/software/dracula/index.html.

6 http://www.francais.ptv.de/ptv.vision/vissim.

2.5. Modeling approaches

Many approaches are used for the simulation of transport systems. A review of simulation approaches can be found in [KIT 05; AKÇ 01].

In this chapter, we present a non-exhaustive list of some of these approaches and references thereto. Other references are cited in the section describing the applications. They allow the reader to deepen their knowledge on the elements presented in this chapter. The approaches listed below are the ones used most widely for the simulation of urban transport systems: Petri networks [BOU 08; DEM 08; TOL 05; DIF 04; NAI 02], multi-agent systems [BAL 08; MEI 07; BAT 05; LOP 08], evolutionary algorithms [MED 08; CHA 95; YIN 02], stigmergic approaches [DAO 08; DAO 06; YU 05], mathematical models [CRA 09; BOU 07; HOU 06; DEY 05; BON 01; GAR 05; AWA 04; PUR 01; LEE 04], cellular automata [CAS 06; ESS 97; BAR 03], fuzzy logic [CHI 02], geographic information systems [HEN 04; BEN 04], object models [CAS 06], and virtual reality [CHU 08].

Finally, models combining several types of modeling [BAR 06] and hybrid models have also been developed. Hybridization of models can be made by considering models corresponding to different levels of modeling, models corresponding to different approaches (multi-agent, Petri, genetic algorithms, Tabu search, etc.) [LÓP 08; KHA 08; FAY 03].

2.6. Fields of application

Here, we will only deal with support applications for live traffic regulation. The complexity of the live control problem is a multi-objective NP-hard optimization problem. Many approaches have been proposed (mathematical models, heuristics and metaheuristics, soft computing, Petri networks and flow simulation).

The control of bus networks has been the subject of numerous studies. The multi-agent approach used by many authors [LAÏ 02] proposes a decision support tool based on a multi-agent approach, but the regulation is only processed at the interchange level; [BAL 08] also uses a multi-agent approach, however, connections are not explicitly addressed; the SATIR (*Système Automatique de Traitement des Incidents en Réseau* – Automatic System for Processing Network Incidents) system was developed by INRETS and Lamsade [BAL 05] for the management of a bus network. [YU 05] uses a stigmergic approach and [CHA 95] uses genetic algorithms to handle problems of correspondence between lines. For the regulation of a bus line, [LI 91] developed a stochastic linear programming model, [BIE 02] and [MOH 98] used evolutionary algorithms. [HUI 02] developed a scheduling approach based on dynamic programming with relaxation. Its goal is to allocate different races to cars for a fixed horizon based on a multi-agent approach. To model connections, [ABB 02] used the stochastic Petri networks. In the same regard, [NAI 02] also used Petri networks associated with the Max-Plus algebra.

The objective is to prevent and resolve conflicts arising from simultaneous occupancy of lanes. [SAL 97] used genetic algorithms for the scheduling of trains, taking into account environmental constraints. More recently (in 2007), Alstom Transport and Eurodécision developed the LP-CDS system as a tool for the detection and management of conflicts, based on linear programming techniques and mathematical models.

Moreover, the psycho-ergonomic aspects were also addressed to characterize the control task, especially for control of the subway. We quote the SAGIM system, designed for RATP[7] for managing incidents in the Parisian subway [ZAN 03] and for using ergonomic recommendations in the development of a support tool to regulate subway traffic [PAS 02].

Several authors have adopted a hybrid approach. [KAC 02] combined blurry logic and evolutionary algorithms. [PAR 05] used genetic algorithms and simulation in the case of a bus network. For the same type of problem, [NGY 07] used simulated annealing and

7 *Régie Autonome des Transports Parisiens* – Parisian transport operator.

genetic algorithms. [FAY 03] coupled evolutionary algorithms with a multi-agent approach in the case of a multimodal transport network.

In addition, awareness of environmental constraints, the necessity for a sustainable development and economic constraints provide a glimpse into the medium term of a significant evolution of urban modes of transport and the appearance of modes of transport complementary to those which already exist [THÉ 03]. These modes range from the logic of purely individual transport (taxi, private vehicle, etc.) to public transport with rigid contours offering a number of flexibility and responsiveness characteristics.

Several research and experimental projects have been conducted or are underway on this topic [SKY 08; ULT 08; BEN 06; ISS 05; MOB 05; CYB 04; JAY 03]. For example, fleets of autonomous or semi-guided vehicles [CHA 04], car-sharing systems, will ensure on-demand customer travel on unfrozen routes. "Platooning" (running vehicles without contact) allows the ambition for an integral and integrated mobility to be transposed into a more tangible reality. Operators of public transport, complementary to existing public transport, will manage fleets distributed in urban areas to provide multimodal fully self-service offers, without reservation. However, these vehicles can be reconfigured into public transport on certain routes with high demand, and at certain times, or in case of malfunction of usual public transport. If, in addition, we associate car-pooling with this reconfiguration, we then have temporary high capacity public transport in case of disruptions to existing public transport or during peak hours, which can become individual on-demand transport when transport demand returns to normal.

The study of multimodal systems is more complex, more recent and still underdeveloped. Simulation is an element of decision support systems for control operators. The SART (*Système d'Aide à la Régulation de Systèmes de Transports Multimodaux* – Support System for Control of Multimodal Transport Systems) project is driven by several research teams (INRETS, LAMIH, LAGIS) and the Transvilles company, allowing the design and validation of a support system prototype to regulate a bimodal tram-bus network [SAR 07]. It was used to evaluate and validate, by simulation, different regulatory models: propagation resolution of blurry constraints [CHI 02], case-based reasoning [BOU 06], evolutionary and multi-criteria algorithms [OUL 06]. [KHA 07] developed a multicriteria itinerary selection approach. Other systems have been developed: [MEL 08; EZZ 08; BAL 07; ARA 04]. However, very little of this work has led to the development of actually operational and marketed support systems.

Many other applications use urban simulation. We cite in particular the development of evacuation plans [PID 96; RAD 03], freight transport [CRA 09; WIS 07; TOI 05], analysis of performance of energy and environmental transport solutions [YAO 06; GUA 98], and intersection management [MED 08; DON 08; DIF 04].

2.7. Software tools

Many simulation tools currently exist, especially for microscopic simulation. The European Commission counted, in 2002 [EUR 02], 57 pieces of software and classified them according to the type of roads considered: urban streets, roads, urban streets and roads, highways). Twelve of them were deemed capable of modeling functions related to ITS [BOX 00] carried out the assessment of simulation tools and identified 65 microscopic models, 16 macroscopic and 3 mesoscopic. They also established a list of models capable of modeling functions related to ITS.

Moreover, as tools become more user-friendly and easy to use, it becomes easier to use several models to achieve a more robust study. For example, in [BLO 00], two models, CORSIM[8] and VISIM were used to improve the quality and reliability of simulation results.

Models and simulation tools are constantly evolving. This trend is particularly oriented towards the integration of several modeling approaches, the development of tools to support the validation of models and to conduct simulation.

8 http://www.kutc.ku.edu.

The next two sections of this chapter present two simulators in the field of multimodal transport and illustrate this trend. The first is a simulator of bimodal guided urban transport systems (tram – bus), developed using the QUEST software tool. The second simulator, based on the principle of self-organization, aims to improve the traffic of individual vehicles. The integration of these two systems will simulate traffic and the control of various types of urban transport: passenger cars, public transport and on-demand transport.

2.8. Simulation of the Valenciennes transport network with QUEST software

2.8.1. *Problem*

The aim of the project is to develop a platform for simulation of the Valenciennes transport network and its interchanges. This multimodal network includes tram and bus lines. Transfers between different modes of transport are provided by interchanges. The interchange located at the railway station provides connections with the arrival/departure of trains.

2.8.2. *Network operation in normal mode*

2.8.2.1. *Bus network configuration*

The bus network consists of 38 lines and three depots located in Saint-Saulve, Denain and Saint-Waast (Figure 2.2).

The lines are of different types (simple lines, X-shaped, with antennas) and consist of a set of itineraries.

By "itineraries", we mean the description of the route of a bus in the network: the itinerary consists of a sequence of points and the description of the itinerary between these points.

A "point" can be a checkpoint which the vehicle is required to pass through or stop at. The different types of stops are:

– “simple” stops: passengers board the vehicle, others alight and leave the network;

– “correspondence” stops where passengers board, others alight to get to another vehicle (bus or tram), and others leave the network;

– “terminus” stops: all passengers alight from the vehicle. Each line contains at least two terminus stops (antenna lines have more).

The lines are regulated according to the “schedule regulation” mode: each stop has a timetable that includes arrival times and departure times of vehicles. Each vehicle is assigned to a line, except reserve vehicles that are parked in the depot and are only used for control operations.

At the start of the day, the vehicles leave the depot to join the terminus of the line and start transporting passengers. Each vehicle is assigned to one route on one line.

There are several types of vehicle (standard, articulated or Midibus). Each of these types contains dimensions (height, length and width), a capacity (maximum number of passengers) and kinetic characteristics (maximum speed, acceleration, deceleration).

2.8.2.2. *Configuration of the tram network*

At first, we were only interested in the first phase of the Transvilles company project, which consisted of implementing a line going from the University to Dutemple. It has 20 stops and the depot is located in Saint-Waast (Figure 2.2). The terminus of the Dutemple and University line can store a reserve train or a broken down train.

The rolling stock consists of 17 trains, characterized by dimensions (length, height and width), capacity (number of passengers) and kinetic characteristics (maximum speed, acceleration, deceleration).

The line is controlled by an “interval regulation” mode to ensure consistency of passage at stops: the frequency of the tram is every five minutes during peak hours and every 10 minutes off-peak.

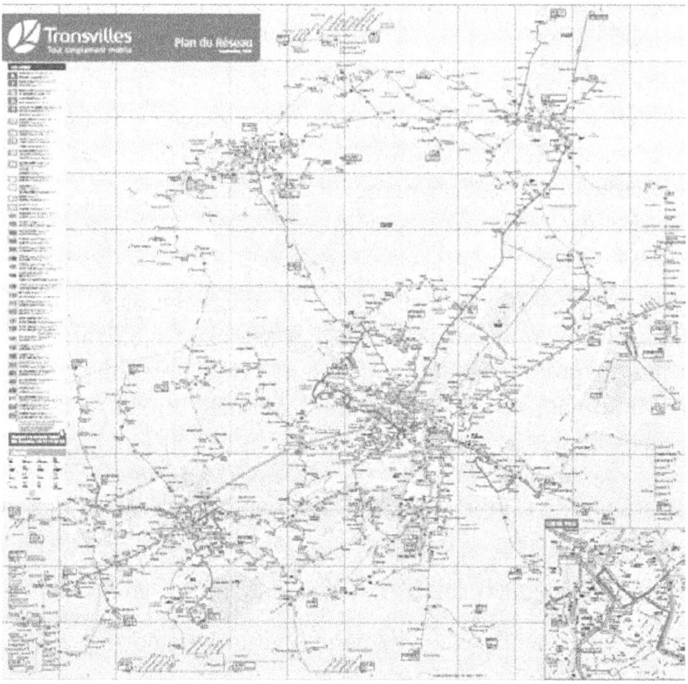

Figure 2.2. *The Valenciennes bimodal network*

2.8.2.3. *Bus/tram interchanges*

The tram network has bus/tram interchanges located at Anzin, Saint-Waast, the University and the train station. There are two types of connection:

– bus/tram connections where the bus waits for the tram;

– bus/bus connections where when one vehicle is expected, the other waits.

2.8.2.4. *Staff*

When a vehicle is to leave the depot, it is assigned a driver. Reserve staff are only used for control maneuvers. The operator suggested that in the simulation, it was not necessary to consider the management of drivers, except for reserve personnel.

2.8.3. *Disturbed mode network function*

The operation of the transport system in disturbed mode is characterized by disturbances that can occur on bus/tram/train networks, and dispatching of bus and tram network regulation procedures which are implemented by regulators.

2.8.3.1. *Disturbances*

The typology of incidents that may occur on the bus and tram networks has been established. Among the main incidents that might affect the operation of a public transport vehicle, we distinguish those that result in non-compliance with scheduled times and those that cause the failure of routes.

Disturbances resulting in non-compliance with schedules are due to:

– traffic problems (peak hours, accident, etc.);

– change in unanticipated demand (cultural or sporting event, etc.);

– the driver (learner, ill, absent, etc.);

– passengers (aggression towards driver, altercation between passengers, etc.);

– technical problems (damaged vehicle, etc.);

– climate conditions (snowy roads, etc.).

Disturbances that cause the failure of routes are due to:

– traffic state (accident, etc.);

– road blocking events (protest, etc.);

– climatic phenomena (icy road, etc.).

These disturbances can cause the advance or delay of a vehicle, vehicle stop or even the non-availability of part of the network.

2.8.3.2. *Regulation procedures for the bus network*

Documents provided by the operator Transvilles define the procedures commonly used for control of the bus network [SEM 04]. These procedures are:

– off-line dead mileage;

– live direct;

– live U-turn;

– U-turn with exchange of drivers and vehicles;

– overtaking and descending service only;

– delayed departure from terminus;

– exchange of drivers and vehicles;

– shortcut;

– on-time delivery and availability;

– line regulation;

– terminus regulation.

Several basic procedures can be combined, in parallel or sequentially, and generate complex actions.

2.8.3.3. *Regulation procedures for the tram network*

The following classification has been synthesized from [FOU 02]:

– exchange at the end of a round;

– dead mileage;

– towing/pushing;

– speed limit;

– withdrawal from reserve dock;

– recovery of service;

– waiting agent.

These basic procedures can be combined to generate complex procedures.

2.9. The QUEST software

To the needs of our application, existing systems dedicated to the simulation of transport networks have the following limitations: the non-inclusion of passengers, the requirement for public transport vehicles to take a fixed path (no possible deviations), and the inability to simulate regulation maneuvers on vehicles.

To be able to simulate a wide variety of phenomena, we retained the QUEST (Queuing Event Simulation Tool) software which we briefly present in the next section.

2.9.1. *Presentation*

DELMIA (http://www.delmia.com), a subsidiary of Dassault Systèmes, develops e-business solutions for the planning, creation, management and control of production systems.

Among these products, QUEST can simulate and analyze the workflow in a virtual 3D environment that provides a realistic, animated, graphical representation and also facilitates understanding, verification and validation of the model. Furthermore, the simulation system can be used by staff (regulators, in our case) without requiring expertise in modeling, programming or simulation.

2.9.2. *Modeling*

The logic model for QUEST is the logical representation of the behavior of the real system. It manages *elements* (active entities that have operation logic, attributes) and *parts* (passive entities that are processed and transformed by *elements*). The *parts* move from *element* to *element* and their behavior depends entirely on the *elements* that process them.

The *elements* and *parts* are defined in QUEST using a methodology based on the concept of class. Each class has a physical representation (3D model), attributes and *logics* that describe decisions made by an *element* at a specific time of the simulation.

Each *element* is associated with a procedure called the *process* that defines the actions to accomplish when a *part* is dealt with by an *element*. We distinguish between different types of *elements*:

– *sources* that create *parts*;

– *sinks* that destroy *parts*;

– *buffers* that store *parts*;

– *machines* that transform *parts*;

– *conveyors* that transport *parts*;

– *labors* that represent human beings involved in the process. They can transport *parts* or perform work on an *element*.

For the transport of *parts* from one point to another in the model, QUEST uses the notion of *AGV* (Automated Guided Vehicle) and involves several types of *elements*:

– *AGVs* are vehicles that move through the system and transport *parts* from one place to another;

– the *controller* supervises and manages the behavior of *AGVs*;

– the *decision points* are the points of loading and unloading of *parts*. They can also be used to manage traffic;

– the *AGV path system* is the network in which the *AGVs* move which also supports the *decision points*.

SCL (Simulation Control Language)

Each *element* of the QUEST logic model has standard *processes* and *logics* for which the user can set parameters through the graphic interface. Often these predefined behaviors do not accurately describe a complex situation and it is necessary to re-define the *processes* and *logics*: in particular, the standard behavior of *AGVs* in their network is very simple and does not permit modeling of the complex operations.

To remedy this, QUEST has a proprietary language called SCL (Simulation Control Language), which can be used to change the behavior of entities in the model by reprogramming *processes'* and *logics'* standards of *elements*.

2.10. Network modeling in normal mode

Modeling is the description of how the system operates according to the software formalism used for the simulation. It requires the use of simplifications of the real system; it is a balance between a fine representation of reality (which requires taking into account a maximum of physical phenomena) and a level of understanding and required accuracy.

Some objects have been modeled easily from one type of QUEST entity and their behavior could be fully defined through the graphic interface. Other objects from the real system were modeled from several QUEST model entities and their complex behavior was the subject of a consequent programming work.

2.10.1. *Topology of traffic networks*

In the application, we distinguish between three networks:

– the road network on which buses will run. In the application, only routes on which the buses are likely to run are modeled;

– the tram rail network. These channels can be optionally supported by a road on which buses run;

– the SNCF network whose modeling is identical to the tram network.

2.10.1.1. *The road network*

These are all the carriageways that may be used by a bus. The road may have one lane (one-way road) or be divided into two lanes corresponding to the two directions of traffic. The set of roads is modeled by a "*path system*". A one-way road is modeled by uni-directional sections placed end to end. Some sections corresponding to two-way traffic (passing under a narrow bridge, for example), may be bi-directional.

2.10.1.2. *The rail network*

Rail networks consist mainly of one-way paths. However, some tram track networks are two-way because they can be used in regulatory procedures. Each rail network is modeled by a *path system* consisting of a set of uni- or bi-directional segments that model the traffic lanes.

2.10.1.3. *Intersections*

It is very important, especially on roads or "road/tram" intersections, to manage intersections in order to avoid collisions between vehicles. The *decision point* can limit the number of vehicles accessing a point. Collisions are avoided by placing a *decision point* at each intersection and assigning it a capacity of one. This QUEST entity, however, fails to manage priorities. Reprogramming of the *decision point* logic is then necessary to promote, for example, the passage of trams.

2.10.1.4. *Overtaking*

– On the rail network

Vehicles on rails cannot overtake each other. The "overtaking maneuvers" are only used within regulatory procedures. They then consist of taking parallel lanes or tracks to maneuver.

– On the road network

Overtaking of a running bus by another bus is not allowed. The only case of authorized overtaking is when a vehicle is at a stop. The stops are modeled by specific clearance lanes, so overtaking of a stopped bus by another bus does not present any difficulties.

2.10.2. *Bus lines*

A line is defined by its set of routes (ordered by sequence of defined scheduled points on an itinerary).

These routes are modeled in QUEST by an ordered sequence of *decision points* on a *path system*. A *decision point* can then be used to describe: a stop point (for the ascent and descent of passengers) or a checkpoint. In this case, management of *decision points* can be complex, if for example it requires directing vehicles to various points based on schedules.

2.10.3. *Vehicles*

The rolling stock consists of two categories of vehicles: trams and buses. These are of various types: there are simple buses, articulated buses, minibuses, etc. that differ in size, kinetic characteristics and capacity.

2.10.3.1. *QUEST entities*

QUEST offers an entity type called *AGV* (Automated Guided Vehicle) that models vehicles moving from point to point within a network. Different types of *AGVs* can move within a network. The standard behavior offered by QUEST is not sufficient to model the behavior of buses and trams but the possibilities for programming behavior with the SCL language allows us to achieve a very satisfactory modeling reality.

2.10.4. *Modeling*

To consider the diversity of vehicles and behaviors, we defined categories for the types of vehicles (tram, train, standard bus, articulated bus, minibus, etc.). For each category, a class (in the QUEST sense) was defined for each business line. Spare vehicles constitute a class.

A vehicle, whether in motion or at rest, has two types of characteristics:

– constant characteristics that are common to all vehicles of the class: identifier, capacity, kinetic limits (speed at empty load, curve at maximum speed, speed at maximum load, acceleration, deceleration), the distance to be maintained between two stationary vehicles;

– individual vehicle variables: position (identifier of the track, position on the track), instantaneous speed, destination, number of passengers.

Each vehicle class also has a physical model that takes into account the vehicle dimensions (length, height and width) and a simplified representation of its shape.

The average speed of bus and trams between two stops is variable. We have integrated the concept of "commercial speed" (average speed between two stops), known by the operator. To consider it, we assigned each stop an attribute that defines, for each type of vehicle, the speed it should take to get to the next stop.

2.10.5. *Stops*

2.10.5.1. *Simple stops*

A "simple stop" is a point where passengers may alight or board without any correspondence. A simple stop has the following characteristics:

– the theoretical arrival time of the vehicle, Harr;

– its theoretical duration, D;

– the passenger arrival rate, Tarr, depending on the time interval and type of day. Tarr is expressed as passengers per minute;

– the rate of passengers descending from the vehicle, Tdesc, depending on the time interval and type of day.

Let N be the number of passengers on board the vehicle arriving at a stop, the number of passengers alighting is then: N x Tdesc and the number of passengers who board is: Tarr x number of minutes since the last visit.

The operation used is as follows:

– if the vehicle arrives on time, travelers alight (number calculated based on Tdesc function); passengers board, limited by places available; vehicle waits time D and leaves;

– if the vehicle arrives early, it waits until the theoretical arrival time, Harr, and applies the previous procedure;

– if the vehicle arrives late before but before time Harr + D, it also applies the first procedure;

– if the vehicle arrives late after time Harr + D, passengers alight; passengers board, limited by places available; and the vehicle leaves immediately.

2.10.5.2. *QUEST modeling*

To consider the above characteristics, we modeled a simple stop by: a *source*, a *machine*, *buffers* and a *decision point* organized as follows:

– the *source* generates *parts* at a rate corresponding to the theoretical time of passage of vehicles;

– the vehicle transports a specific *part* used during stops;

– the *machine* has a *cycle process* of duration D which, in order to be carried out, requires the *part* from the *source* and the *part* from the vehicle;

– upon arrival of the *AGV* to the *decision point*, the *part* is unloaded and used by the *cycle process* of the *machine*. The vehicle remains waiting and the *cycle process* is responsible for changing the number of passengers on board the vehicle and the number of passengers waiting at the stop;

– at the end of the *cycle process*, the *part* from the *source* is destroyed and the vehicle *part* returns to the vehicle.

This organization provides a simple model for the operation described above. The configuration of simple bus and tram stops is shown in Figure 2.3.

2.10.5.3. *Connections*

Connections that combine several modes of transport can be very complex and operate in very different ways. We have divided them into two categories:

– Bus/tram connections: the operation of these connections varies depending on networks, but the most common behavior is the following: the tram passengers alight without worrying about the connection, if necessary the bus waits the connection time and carries passengers.

– Bus/bus connections: again, behaviors are multiple. In most cases, however, a connection is organized around the vehicle being awaited (bringing passengers) and the vehicle waiting for it (to carry passengers).

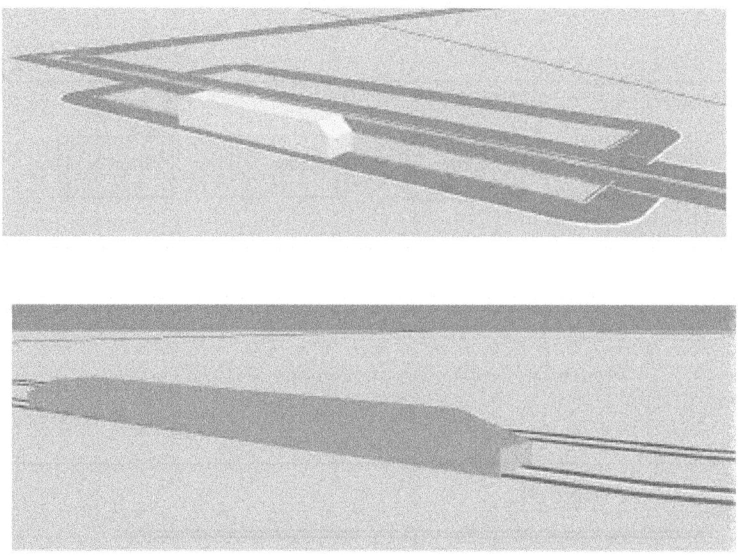

Figure 2.3. *Organization of bus and tram stops: the bus takes a clearance route while the tram continues to occupy the lane*

a) Bus/tram connections

The tram stop is a simple stop that generates an extra *part* for the completion of the connection.

The connection is modeled by a *machine* whose *cycle process* of duration D (theoretical duration at the stop) requires:

– a *part* from the tram which guarantees that at least one tram has passed;

– a *part* from the *source*: this is generated according to connection schedules and ensures the connection implementation at the correct time (or after it, but not before, even if the vehicles are present);

– a *part* generated at the arrival of the bus that only allows the connection to occur if the bus has arrived.

Figure 2.5 illustrates the modeling of this type of connection. It uses the following graphical representation of QUEST entities (Figure 2.4).

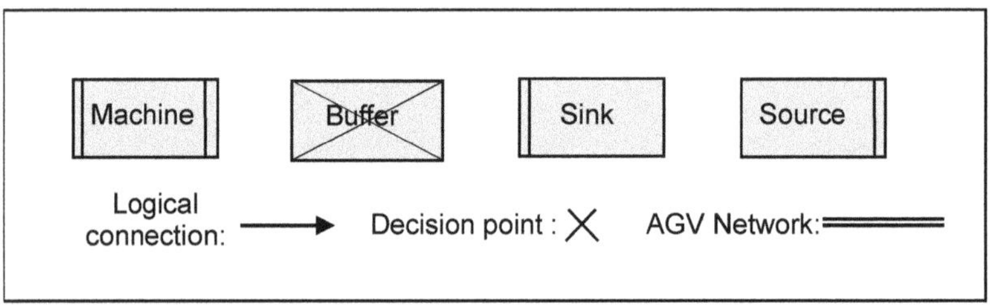

Figure 2.4. *Graphical representation of QUEST entities*

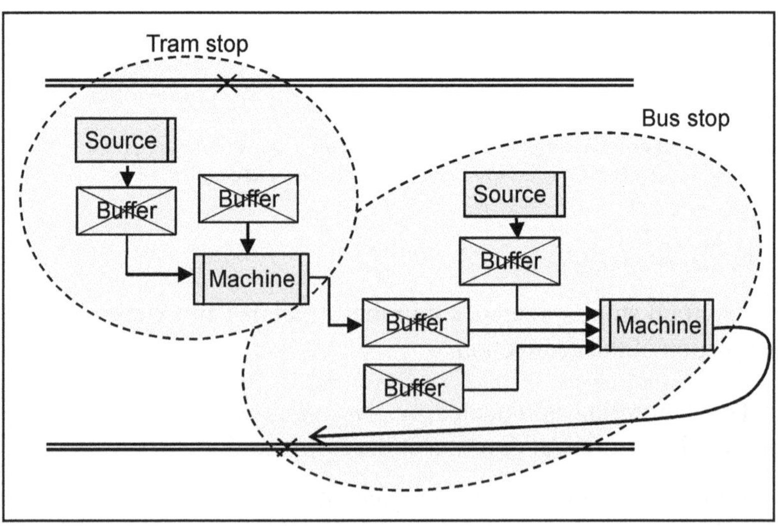

Figure 2.5. *Bus/tram connection model*

b) Bus/bus connections

Modeling a bus/bus connection involves NV vehicles including some in waiting and others being waited for. It consists of two steps (Figure 2.6):

– A *machine* is powered by *parts* from each of the vehicles participating in the connection. This ensures that each vehicle is presented in the connection. The duration of the *cycle process* is zero.

– The previous *process* generates NV *parts* that are sent to NV *machines*. These *machines* have a *cycle process* similar to that of a simple stop: it requires, in addition to the previous *part*, a *part* generated by a *source* at a time corresponding to the arrival of the vehicle and having a duration corresponding to the waiting time of the vehicle in the station.

2.10.6. *Passengers*

From an origin/destination survey conducted on the network before its reorganization, we established a correlation between the current and previous configurations and we assumed that the day could be divided into time periods (peak hours, off-peak hours) during which the flow passengers are regular. For a given time period, we considered the flow of passengers as stationary and the parameters we considered were:

– at each stop, the average rate of number of passengers arriving per minute;

– for each vehicle at each stop, the rate of descent of the passengers as a percentage;

– for each vehicle at each connection, the percentage of passengers who alight and the percentage of passengers who go to each of the connecting vehicles.

Passenger arrivals at a stop are modeled by a Poisson distribution. Each stop is characterized by an average rate of passengers arriving per unit of time, denoted m. N number of travelers arriving at a stop during time interval Δt is characterized by a random variable with probability density:

$$P_N \Delta t = \frac{(m\Delta t)^N}{N!} e^{-m\Delta t}$$

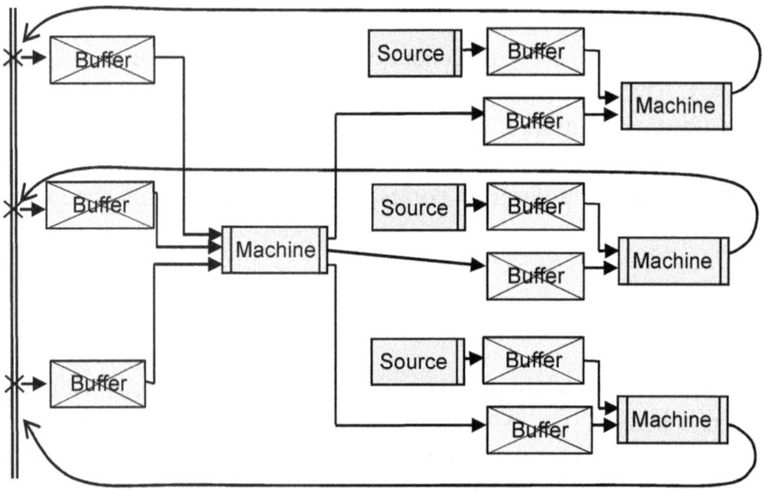

Figure 2.6. *Modeling a bus-bus connection between three vehicles*

Depending on the time of day or the characteristics of the day (working day, holiday, holiday season, etc.), the average arrival rate of N users may be different. This approach has been adapted to each type of stop (simple, terminus, connection).

2.10.7. *The flow of connecting passengers*

For each stop, i, of the connection, the flow of passengers is defined by:

– the passenger arrival rate (passengers per minute) from outside of the network: $Tarr_i$;

– the rate of descent from the vehicle to outside of the network: $Tdesc_i$;

– the rate of passengers leaving vehicle i to get into vehicle j: $Ttransf_{i,j}$.

For vehicle k with N_k passengers on board, when stopped, the following transfer of passengers will be made (in this order):

– N_k x $Tdesc_k$ passengers disembark and leave the network;

– N_k x $Ttrans_{k,j}$ passengers are transferred to the connecting stop, j;

– all passengers waiting for available seats board the vehicle. Passengers waiting include passengers from outside the network and passengers who alighted from another connecting vehicle.

The consideration of this operation is modeled in QUEST through class attributes, representing the stops and vehicles. The evolution of these attributes is provided by logics.

2.11. Network modeling in degraded mode

2.11.1. *Disturbances*

Management of a vehicle traveling on a network is hermetically managed by QUEST. This management is of the *AGV* (Automated Guided Vehicle) type: when several vehicles are moving on the same network segment, the vehicles are maintained behind each other, regardless of their individual speed. It follows that if a vehicle stops or slows down, the following vehicles are consequently blocked. This travel management is consistent with movements on a guided tram network, but is not suitable for bus management.

To overcome this problem, it was necessary to develop clearance areas all along the road network frequented by buses, to hold broken down vehicles thus allowing subsequent vehicles to overtake. These clearance areas were arranged near and between bus stops. To ensure consistency in the management of disturbances, on the guided tram network we defined similar clearance areas for receiving trams subjected to a disturbance (Figure 2.7).

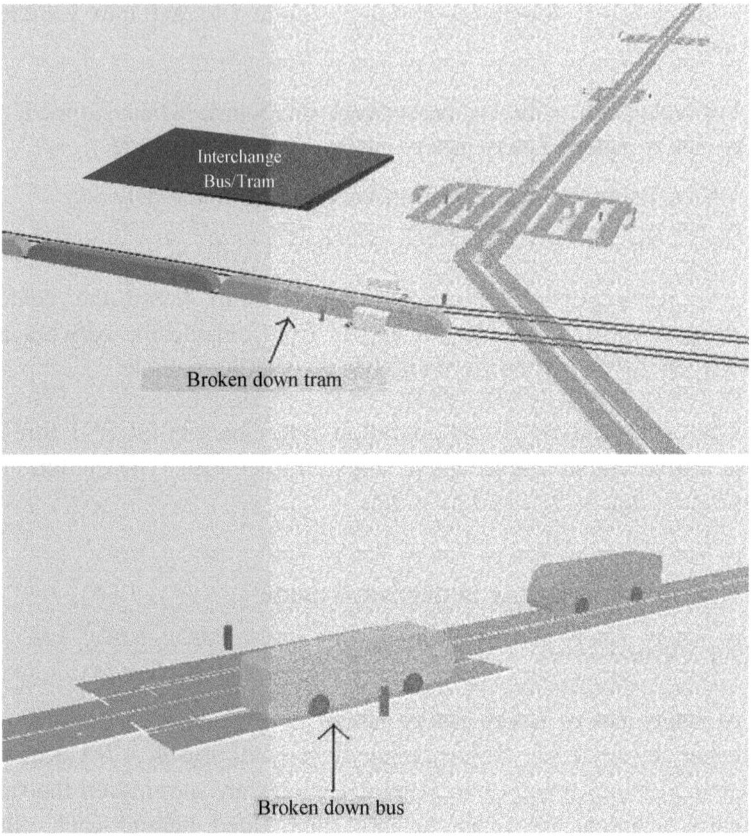

Figure 2.7. *The clearance areas for buses and trams.*
Top: the broken down train blocking the flow of the following trains.
Bottom: the broken down bus parks in a clearance area

A disturbance is generated in the following manner:

– To generate a delay, the user designates a vehicle and indicates the delay t to be applied. The vehicle goes to the next clearance area where it waits time t then resumes its normal route. The control support system (*Système d'Aide à la Régulation* – SAR) is then notified of the vehicle delay.

– In case of stoppage: the user designates the vehicle and indicates duration t at the stop to be applied (optional because the duration of a breakdown is not predictable), the vehicle goes to the next clearance area and stops there. The SAR is then notified of the vehicle that has stopped.

The early arrival of a vehicle does not constitute a disturbance because, according to what happens in reality, early arrivals are absorbed during stops.

2.11.2. *Regulatory procedures*

The regulatory procedures presented above were decomposed into a basic set of actions. The simulation of a regulatory procedure decided by the SAR is therefore to simulate the basic actions in parallel or sequentially.

As an example, we present the basic actions that make up the programming of a "live direct" regulation action.

Example: live direct:

– V_r is the delayed vehicle, V_{r+1} the following vehicle on the same line, V_{r-1} the preceding vehicle on the same line;

– R is the delay of vehicle V_r relative to the theoretical schedule table TM;

– T2 is the terminus of the line in the current direction (line direction that V_r and V_{r+1} are running in), T1 being the terminus of the line in the other direction.

Conditions of application: the delay R of vehicle V_r is such that V_r was joined by V_{r+1}.

Live direct consists of achieving a transfer of passengers from V_r to V_{r+1}. V_r is sent directly to the terminus T1 by U-turn and then takes up the TM schedule. V_{r+1} keeps its normal schedule until T2, is advanced by R/2 at T2 and resumes schedule TM at T1. V_{r-1} is delayed by R/2 at T2 and resumes schedule TM at T1 (Figure 2.8).

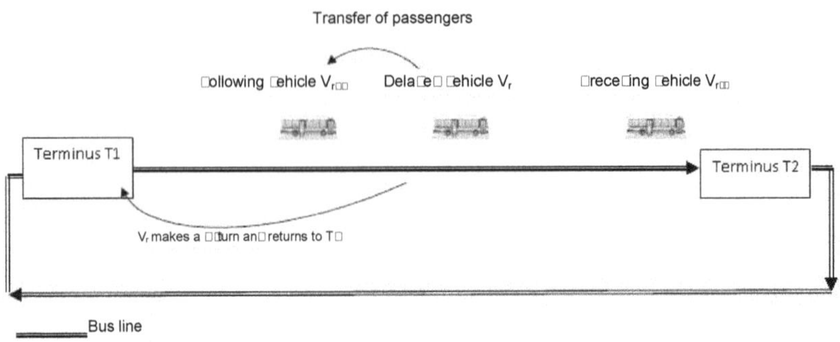

Figure 2.8. *Live direct*

Assuming that the transfer of passengers takes place at stop n, this regulatory procedure results, in QUEST, in programming the following basic actions:

Let: t_{ret}, be the date that the delay was processed; t_k^1, the date of arrival of vehicle V_k at T1; t_k^2 the date of arrival of vehicle V_k at T2.

At t_{ret}:

– transfer passengers from Vr to Vr+1;

– send Vr to T1 with sufficient speed so that it arrives before the next scheduled stop;

– continue Vr+1 normally (with its planned schedule) until T2;

– continue Vr-1 normally (with its planned schedule) until T2;

– remove time of passage of Vr between stops n and T2, and between T2 and T1.

At tr-12: V_{r-1} arrives at T2, delay it by R/2, i.e. delay its scheduled time at all stops between T2 and T1 (without changing its speed).

At tr+12: V_{r+1} at T2, bring it forward by R/2, i.e. bring forward its scheduled time at all the stops between T2 and T1 (without changing its speed).

At tk1 for k = r-1, r, r +1: the vehicle Vk returns to its normal schedule.

2.12. Simulation results

Apart from results typically provided by a replication (trace report, summary report, etc.), the results provided by the simulation system concern the evolution of indicators that we have divided into two types:

– indicators that relate to the service provided to users;

– indicators relating to vehicle use.

The indicators that are involved in the provided service are:

– waiting times of passengers at stops and in connection;

– number of passengers waiting at stops;

– rate of filling of vehicles;

– delays or early arrivals of vehicles.

Those that involve the use of vehicles are:

– speed of vehicles;

– circulation time and waiting time;

– delays and early arrivals;

– irregularities in frequency of vehicle passage;

– number of passengers transported.

2.13. Conclusion/perspectives

The above results have all been implemented within the QUEST software. The model developed was validated using different scenarios. We have endeavored to build a user-friendly interface operable by a non-specialist user of the simulation.

On the other hand, we tried to develop a fairly realistic 3D physical model of the network which nonetheless reflects the compromise between the smoothness of 3D modeling and animation speed.

However, extending the functionality remains subject to the "opening" of QUEST. In this regard, QUEST has some shortcomings, namely the very laborious nature of the definition of a transport network, insufficient traffic rules, failure to integrate specific vehicles.

Moreover, the emergence of the problem of on-demand transport, coupled with the regulation of multimodal transport networks, has led us to undertake the development of a simulation system based on a traffic simulation module developed within the team (FORESEE, see next section). This system must be capable of simulating the flow of private vehicles following realistic traffic rules, traffic control and transport vehicles (buses, trams), management of on-demand vehicles and their interaction with fleets of transport vehicles.

2.14. Self-organization of traffic – the FORESEE simulator

2.14.1. *General problem*

2.14.1.1. *The context*

The development of transport in both urban and suburban or inter-urban zones poses many problems in most industrialized countries. One of these is the chronic formation of traffic jams that constitute a colossal waste in ecological and economic terms and quality of life. These traffic jams often have a random nature which makes them impossible to proactively manage by planning techniques.

Our study is in this context: it aims to improve the traffic flow of individual vehicles through the development of a navigational aid system based on the early detection of traffic.

We will present a novel strategy against congestion which is based on the self-organization of traffic. We will then describe a simulation platform to test this strategy, defining the application modalities (that depend on a large number of parameters) and limits. Finally, we will present the first results of this research.

2.14.1.2. *Centralized organization vs. self-organization*

Two types of approach have been studied and/or implemented to improve the quality of traffic in this context:

– The *centralized approach* is based on a client-server architecture to make traffic information available to users[9]. This approach has the advantage of being conceptually simple but requires the implementation of two infrastructures:

- an infrastructure for collecting traffic information. These are often automatic systems based on video cameras or simply human observations. This can only be considered in a limited context (e.g. urban) and leads to the scalability of the solution. More recently, we used the vehicles themselves as collection agents [DAS; LOR 05];

- a diffusion infrastructure. This is already available (and is usually the mobile phone network). Technically, such a huge operation of the telephone network requires a specific design.

These infrastructures represent a significant cost to the user, especially in the case of permanent connection. Furthermore, the provision of information is subject to a subscription representing an additional charge[10].

As the update of centralized information occurring at time intervals is up to several tens of minutes, the reactivity of these systems is not great.

Finally, the mass diffusion of the same information can cause instabilities leading to the formation of traffic jams as large as those it is supposed to prevent. This kind of phenomenon has been observed for example in Japan in very dense urban environments where this type of organization has been implemented [SHI 99; FUK 02].

– The second approach is based on *self-organization of traffic* where each vehicle can be considered as a sensor and/or consumer of

9 Tom-Tom traffic, Dash Navigation (GPRS), TMC (Radio-FM).

10 This cost is about 120 euros/year for currently available systems.

information. The principle is "word of mouth": it is a cooperative approach. Each vehicle is informed by anonymous partners (other vehicles) about the state of the transport network, and reorganizes its path accordingly [GIB 09; SCH 07; NAD 04; WIS 04].

In this model, the global state of road traffic is not known, only *local vehicle views*. These views are potentially different, and may also be inaccurate.

The implementation of this approach is based on two technologies: GPS geo-positioning and radio communications between vehicles. These technologies currently diffuse massively and represent a very low investment cost. The operating cost of the cooperative approach is zero, which is an important advantage compared to the previous solution.

The non-uniqueness of local views of traffic is a source of diversity in the responses of each vehicle to the problems that arise. This makes the instability phenomena already mentioned very unlikely.

The diagram in Figure 2.9 shows the structure of the embedded system in each vehicle. It comprises a conventional navigation system, a GPS receiver and a bi-directional communication unit. A filter/aggregation module builds a local view of the traffic situation.

2.14.1.3. *The challenges of self-organization*

Despite its qualities, the cooperative approach poses a number of implementation problems.

The first concerns medium range communications: technology supports exist, but the choice of one of them bears no clear consensus. For economic reasons related to large-scale production, and technological reasons related to their performance, products conforming to the IEEE 802.11a/b/g standard (known as Wi-Fi) seem to be good candidates.

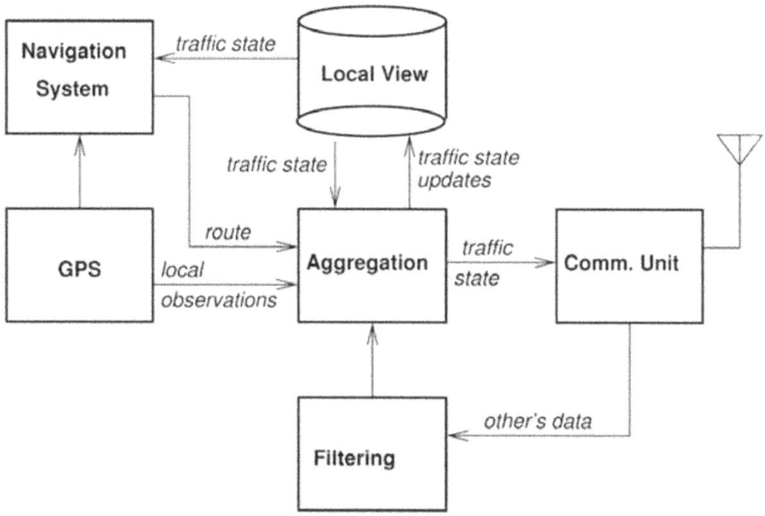

Figure 2.9. *Structure of the embedded system*

However, one must be aware that the use of this media may conflict with the classical use of Wi-Fi, especially in urban areas. For this reason, a variant of the classical 802.11 standard, called 802.11p appeared [TIB 07]. It is dedicated to uses in which mobility is greater and has the following advantages: different frequency band than previous wireless standards, non-recovery of channels and higher range [RUB 07].

Paradoxically, one of the problems posed by this type of communication is not necessarily the lack of communication, but perhaps its excess [ART 07; TOR 07]. This can happen in congested traffic, where each player tries to inform others. This is called *informational storm.* This phenomenon generally affects communications by masking minor but significant information.

However, the real challenge of self-organization is mostly logical order: each player in traffic must build their own picture of the state of the road network from the information they have collected themselves, but also those received from other players. This image will allow them to anticipate problems located on their itinerary.

This implies that the information provided by each vehicle must be relevant (that is to say useful to others). For the reasons discussed above, the amount of disseminated information must be controlled, thereby limiting the risk of *informational storms*.

Aggregation algorithms must therefore be developed, taking into account:

– the existence of similar information from different sources;

– the existence of contradictory information;

– the temporal relevance of the information (which becomes less reliable with age);

– the geographical relevance of information.

2.14.1.4. *The aim of the simulation*

Self-organized traffic can be seen as a system with a large number of players operating in a diverse environment. Interactions between them are numerous. They can be:

– physical (related to the movement of vehicles);

– logical (related to exchange of information between vehicles).

The parameters that influence these interactions are also numerous:

– nominal capacity of the road network (speed limit, number of lanes, etc.);

– various disturbances (weather, accidents, etc.);

– user needs in terms of transport;

– individual driver behavior;

– type of communication media used;

– influence of the environment on communications and geolocation (urban areas, tunnels, etc.).

Finally, the criteria for quantifying the performance of the system are varied:

– temporal (minimization of travel time);

– energy (minimization of fuel consumption);

– ecological (avoidance of poorly ventilated areas during pollution periods);

– equity (avoidance of excessive penalization of a certain category of users by optimizing a global criterion).

This diversity – linked with the difficulty of establishing field experiments – makes it almost mandatorily a simulation study. We will now present the simulation platform that we have developed to meet this requirement.

2.14.2. *FORESEE simulator*

2.14.2.1. *Definition and objectives*

FORESEE is a simulator dedicated to the study of a self-organizing traffic. It is a discrete event simulator, with a graphical interface to monitor and control the evolution of traffic and inter-vehicle communications (Figure 2.10).

The purpose of this tool is to evaluate the effects of different parameters that influence the quality of traffic, but also to identify the best means (tools and strategies) to improve its quality. It provides qualitative (direct observation of traffic) or quantitative information (keeping a log that can be used later).

2.14.2.2. *Simulator duality*

Our tool must simulate two types of flow: the vehicles (physical traffic) and the information (inter-vehicle communications). These two flows are subject to very different laws and evolve according to remote time constants of several orders of magnitude.

The simulator includes a traffic simulation module and a communications' simulation module. They work together under the control of an application module.

This integration is one of the strengths of our system as it enables a *holistic* approach to problems. Many previous studies used a general traffic simulator to generate a traffic scenario (which is static), and a second simulator, completely separate from the first, to simulate communications (often simulators developed to study information networks) [SAI 04; ART 04; MAS 05; GRO 06]. Such an approach allows only one-way interaction between the two simulators.

However, a characteristic of the studied system is the feedback effect operated by communications on flows of vehicles. It is therefore important that the two simulators can run simultaneously and thus interact [SHI 08].

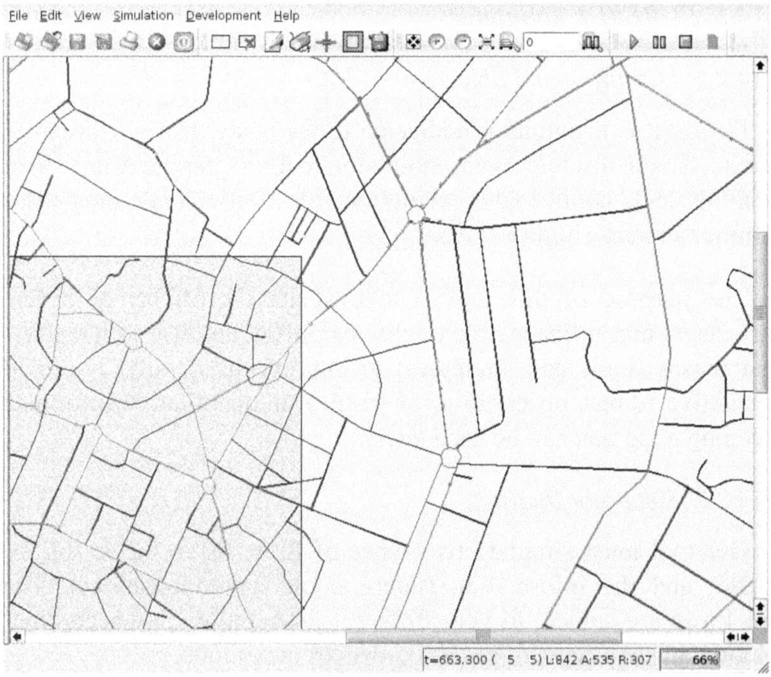

Figure 2.10. *Graphic interface of the FORESEE simulator*

2.14.2.3. *Road traffic simulator*

The road traffic simulator is a discrete simulator with temporal sampling (time-slicing). It manages a collection of mobile agents (vehicles) organized in files on different axes of the network. The sampling period must be consistent with the movement of vehicles, but also the events induced by the communications simulator. A period of 0.1 s to 0.5 s is generally suitable for this type of study.

The *traffic model* defines the behavior of mobile agents. This model takes into account basic traffic rules: behavior at intersections, roundabouts, etc. For the proposed study, it is also necessary that it can generate (and reproduce) settings that lead to situations typical of a perturbed system, such as queues, slow-moving "accordion" traffic. Finally, this simulator must be able to work on a realistic road network model, for example based on actual cartographic data.

The second component of the simulator is a *vehicle tracking model*, which takes into account the situation of each vehicle, its geometric and dynamic characteristics, as well as the characteristics of the driver. This model is based on the acceleration (positive and negative) allowable by the vehicle, a safety distance depending on the speed, a nominal speed, and driver reaction time [TRE 00].

2.14.2.4. *Communications simulator*

For a real implementation, FORESEE uses the IEEE 802.11x specifications, known as Wi-Fi. The advantage of using readily available technology is to be able to calibrate our simulation model by doing field experiments using existing equipment.

The communications' simulator is seen by the traffic simulator as an embedded module that performs the following tasks at each simulation:

– store messages (from each vehicle) to be transmitted in the next time slice;

– establish a schedule determining an initial (random) order of message transmission;

– run a micro-simulator managing the routing of messages to the receivers. This includes the allocation of channels with a CSMA/CA[11] protocol [GER 06]. Messages that have not been sent are re-sequenced to the earliest, but with priority favoring the most recent messages. Messages that are too old are simply "forgotten" after a configurable time;

– run a micro-simulator managing the reception of messages and elimination of packets noisied by collisions caused by the phenomenon of hidden callers (hidden terminal effect) [MOH 00].

The communications simulator is totally different from the traffic simulator, since it is based on a schedule (not a splitting of time into equal slices).

2.14.2.5. *Application module*

This couples data from the two previous modules and simulates the computer system on board the vehicle. Its main role is to detect congestion states and update a local view of the traffic network.

Schematically, the congestion detection algorithm compares the legal speed of the network with the current vehicle speed, and calculates an instantaneous quality indicator. After filtering, a congestion indicator allows the system to switch from normal mode to congestion mode, and vice versa. In case of congestion, the vehicle periodically retransmits its position and the instant quality to which it corresponds.

Receiver vehicles aggregate the qualities received and calculate, for each road segment, an average quality in a sliding time window in which the samples are weighted by the inverse of their age. In the absence of information on a given stretch of road, it is considered that the quality of this stretch is nominal. As a result, the average quality rises progressively after traffic until it reaches the nominal quality. To reduce the memory of average qualities, redundant information is deleted.

---

11 Carrier Sense Multiple Access with Collision Avoidance: access to media by listening to a carrier, with collision avoidance.

2.14.2.6. *Results' analysis tool*

Visual observation of the simulation provides a qualitative estimate of the influence of parameters of current simulations. These parameters are grouped into scenarios, exploring other possible scenarios to better understand the phenomena that enter into play.

A quantitative analysis is needed to detect trends and especially to make objective results. To do this, the simulator is equipped with *monitors* that record variables of interest in a database. A screening tool, separate from the simulator, can extract desired information from the database in order to generate diagrams (such as those illustrated later in this chapter) putting the phenomena in evidence.

2.14.3. *Results*

2.14.3.1. *Evaluation of the effects of self-organization*

The first experiments conducted with our simulator focused on scenarios of urban and suburban traffic in a small town. We set the vehicle flow in the scenarios to obtain an operation at the congestion limits of the network. A different task is assigned to each vehicle (couple start/finish point). These tasks correspond to the traffic flow that we wish to study (work–home journeys, for example). The maximum speed of vehicles corresponds to the maximum speed allowable on the route taken by the vehicle, with a random dispersion of +/-20%. This last point takes into account individual driver behavior, causing them to run faster or slower – potentially beyond the legal speed.

The first set of measures that were carried out correspond to an open loop operation: each vehicle calculates the route best suited to its mission, based on the criterion of shortest duration. This itinerary is followed regardless of any other information.

The second experiment implements the same conditions but the vehicles exchange information on the state of the network. Exchanges take place within a limited range (about 200 m). Every 15 s, each vehicle will calculate a route starting at its current position, based on its own view of the network. The new route will be implemented at the next intersection, when appropriate[12].

In both cases, we calculate a performance indicator which is the quotient of the theoretical travel time (without any traffic jam) over time measured by simulation.

The two experiments last 5,000 s. In the first experiment, many queues form, greatly disrupting traffic. In the second, the number and severity of traffic is less. A visual examination during the simulation clearly shows the bypass measures taken by vehicles to avoid a congested area.

The performance indicator confirms these observations: in the first experiment, it tends to degrade over time, with an average of about 20%. However, in the second experiment, it is stable in steady state, with an average of about 65%, without going below 40% (Figures 2.11 and 2.12).

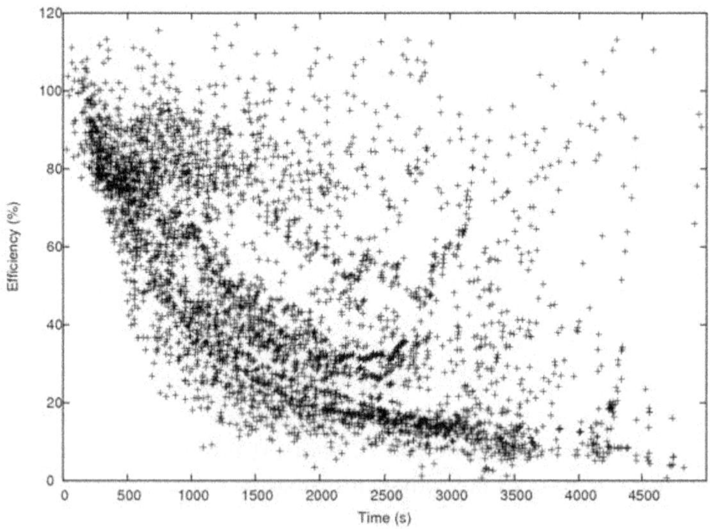

Figure 2.11. *Traffic performance (static routing, according to nominal capacities)*

12 The traffic simulator only allows U-turns as the road graph nodes.

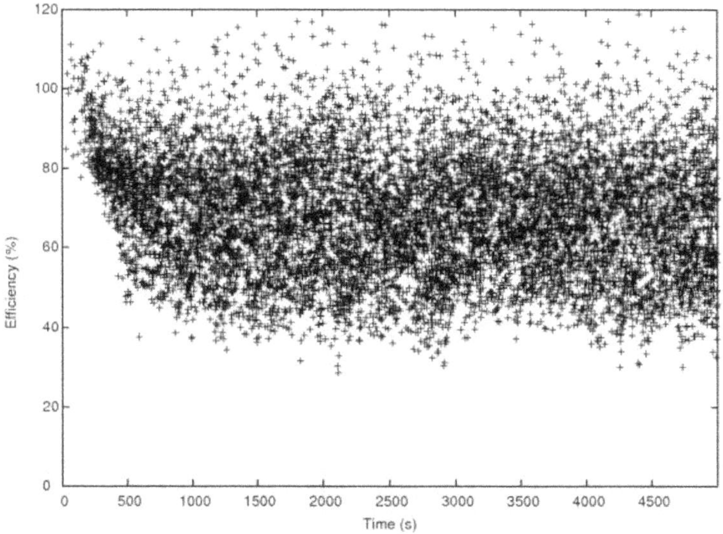

Figure 2.12. *Traffic performance (dynamic routing, based on local view)*

The number of vehicles that arrived at their destination at the end of the simulation confirms previous steps: in the first case, only 50% of 11,049 vehicles arrived. In the second case, there were 90%.

This type of experiment shows two results. First, it is possible to efficiently disseminate traffic information. Secondly, this information, although partial, is sufficient to improve traffic significantly.

These results are qualitatively convincing, but the influence of the model parameters that were used to obtain them is currently not well known. We will thus present another study to determine an effective value for the parameters of this model.

2.14.3.2. *Evaluation of the influence of radio communications range*

To evaluate the influence of the range of radio broadcast, we measured its impact over the proportion of vehicles that receive a *useful* message. "*Useful*" here signifies that this message includes information on the route planned by the receiver vehicle.

This type of information is an evaluation of the quality of interactions, and is very difficult to achieve without a simulator because it depends on traffic conditions at any time.

We set the radio communication simulator to reflect the urban environment:

– Nakagami Mitigation Model [RUB 07; SCH 07]. This model provides the probability of receiving a radio signal depending on distance. It is possible to set it to represent the best performances of the communication system in each type of environment (urban, highway, countryside, etc.). For this experiment, we choose a form factor of one, said to be adapted to the urban environment.

– Directional broadcasting of the signal. Broadcasting of radio waves in urban areas is very constrained by the presence of buildings [RUB 07]. This is a complex phenomenon, but as a first approximation, diffusion along road axes is by far the most effective. To reflect this, our broadcasting model implements the notion of oriented diffusion, focusing on roads. For this experiment, the quality of broadcasting perpendicular to roads is only 1/10 of the nominal quality.

– Only four channels are used, with a flow rate of 54 Mb/s. The Wi-Fi standard currently has 14 channels, but the corresponding frequency bands overlap. Use of just three or four channels ensures no interference. It also allows for less impact on the traditional use of Wi-Fi communications.

– All vehicles are considered to be equipped.

We studied the behavior of the communications system whose ranges are 25, 50, 100, 200 and 500 m. The first case is very pessimistic compared to the performance of standard communication systems, while the latter is rather optimistic.

The diagrams in Figures 2.13, 2.14 and 2.15 show reception rates for ranges of 25, 100 and 500 m respectively. Each measurement corresponds to a message. The "*message identifier*" axis shows the number of messages, but can also be seen as an (irregular) time axis because the numbers are assigned to messages in chronological order.

Figure 2.13 shows that a range of 25 m is insufficient (even in an urban environment) because the vast majority of useful messages are not received. In Figure 2.15 (100 m), one can note a significant improvement in performance, which is confirmed by Figure 2.14 (500 m). In the latter case, the rate of reception is almost perfect, since it varies between 90 and 100%.

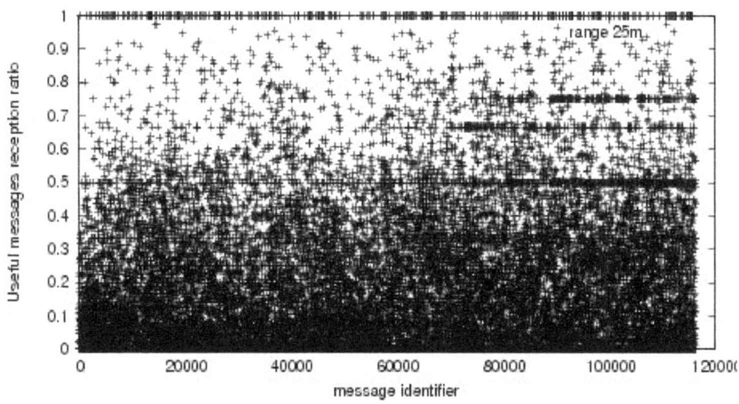

Figure 2.13. *Rate of incoming messages with a range of 25 m*

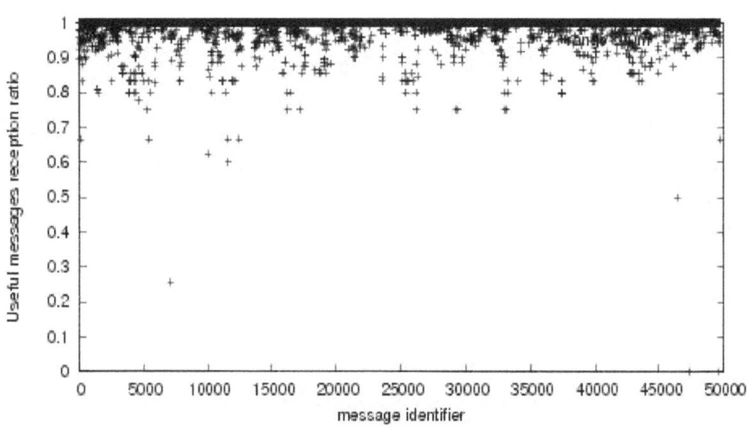

Figure 2.14. *Rate of incoming messages with a range of 500 m*

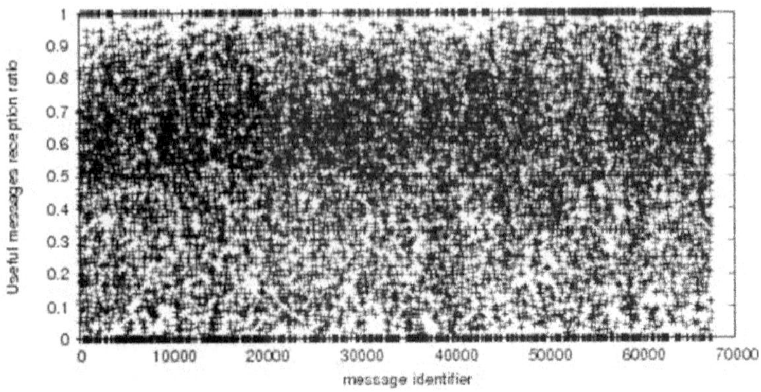

Figure 2.15. *Rate of incoming messages with a range of 100 m*

This visual analysis can be refined (and quantified) by the diagram in Figure 2.16. It is the representation of a probability density estimator for receiving useful messages, calculated from previous data for each of the five cases studied.

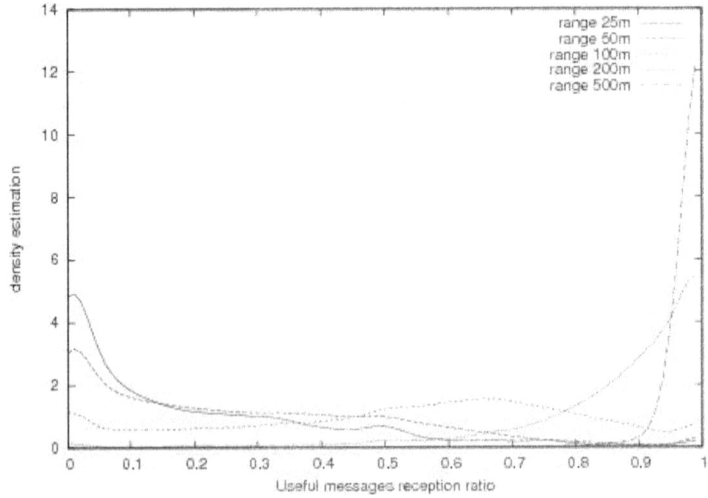

Figure 2.16. *Density probability estimate for receiving different radio ranges*

One can, for example, see the poor performance with a range of 25 m (reception rate < 0.5 for almost all samples) and very good performances in a range of 500 m.

The most interesting case appears with a range of 100 m, for which the probability density is remarkably constant. This is the limit below which the system behavior can be described as "pretty bad".

Examination of Figure 2.16 suggests that the performance of the system with a radio range of 500 m is much better than if it were only 200 m.

Table 2.1 puts these results into perspective. It provides an assessment of the final quality of traffic in two ways:

– by comparing the number of vehicles that arrived to the total number of vehicles that were in circulation;

– by specifying the number of messages exchanged. This number is a direct marker of traffic since normal traffic flow does not give rise to any transmissions.

It becomes apparent that the quality increases rapidly when the range goes from 25 m to 50 m, then to 100 m and 200 m, but it varies little between 200 m and 500 m. It therefore seems unnecessary to use a range of 500 m in the conditions of the experiment.

This type of result clearly demonstrates the relevance of a simulation in solving complex problems involving a large number of parameters. We note that a drastic improvement in the rate of useful messages received (flattering as it is) is not necessarily equivalent to an improved final result. In other words, a simulation study not only helps to optimize a part of a problem, but also to assess the impact of this element on the result (influence and field of usable values).

Range	Vehicles arrived/left	Number of transmitted messages
25 m	6,192 / 7,956	116,396
50 m	6,700 / 7,956	80,559
100 m	6,879 / 7,956	67,449
200 m	6,954 / 7,956	51,618
500 m	7,013 / 7,956	49,756

Table 2.1. *Final quality traffic based on the communication range*

2.15. Conclusion – perspectives

The implementation of self-organization of traffic in a cooperative approach can only be done after a simulation validation phase. We developed it for a specific simulator, whose originality is to combine a physical traffic simulator with an informational traffic simulator.

The first results provided by the simulator show the validity of the cooperative approach in an urban environment. They were obtained using an exchange protocol for simple data, based on the dissemination of information, and immediate retransmission by the most remote receivers. Despite the simplicity, this approach works well in urban and suburban areas, and is compatible with the communication distances provided by the available equipment.

Outside of this context, we will see that a number of problems remain to be solved. All of these problems are the substance of our future research.

2.15.1. *Sustainability of the information*

It is likely that in inter-urban areas, the simple protocol used to validate our approach in an urban environment will not be effective. Discontinuities in the flow of vehicles that can be observed so often induce breaks in the flow of information. This was clearly demonstrated in the second study when the communication range was very weak. This problem is specifically studied in [THO 12], where several solutions are proposed.

In this case, messages are "lost", but they would be useful to more distant vehicles. To counteract this phenomenon, each vehicle must carry information in a physical way, in order to transmit during the period in which it is valid.

2.15.2. *Information aggregation algorithms*

To be stored and transmitted in an efficient manner, information must be aggregated. However, its relevance must always be known in order to avoid spreading unreliable information. This reliability can be measured using the information age, origin (measured or received), reliability (multiple sources), etc.

2.15.3. *Cooperation efficiency*

Much of the information known by a vehicle is not directly useful to it. Conversely, the information that would be useful may be held by another vehicle. Therefore it would seem logical that each vehicle distribute queries containing its route, so that a partner vehicle can disseminate relevant information. It is a form of "a-minima" of two-way communication that is not necessary for urban trips but could be essential in the case of long trips containing "critical areas" such as crossing big cities.

2.15.4. *Deployment of the proposed approach*

As in any cooperative approach, proper operation is only possible by obtaining the support of the largest possible number of partners. This is more of a socio-economical (or political) issue than a technical one. However, it is nonetheless possible to provide some technical solutions to begin the process of cooperation.

The first is to provide the system with very cheap (even free) software and run it on hardware that is already used for another purpose. The technical constraint here is to use hardware with a GPS receiver and Wi-Fi interface[13].

The second possibility is to use public willpower to deploy the first cooperative customers (bus equipment, taxi equipment or fixed terminals used as relays). It is likely that the first implementations will be in urban areas.

Also in this domain, a simulation study can provide valuable information on the level of equipment that would qualify for an interesting operation.

For those who doubt the feasibility of such a deployment, there is at least one example of a cooperative system that has been successfully deployed from scratch: the FLARM system. FLARM[14] is an anti-collision device created by Swiss gliders to improve the safety of flight in a glider. It also uses GPS and radio technology, and is of course only effective between two equipped devices. In just a few years, more than 13,000 pieces of equipment have been installed on gliders, tourism aircraft or microlights (mostly in Europe), which is remarkable for such a small market.

The reasons for this success are in part due to a clear desire from the users, the quality of service and also the moderate cost (equivalent to a pair of seat belts).

---

13 This is the case of some "smartphones" and some "PNA" (Personal Navigation Assistant).

14 http://www.flarm.com/.

2.16. Bibliography

[ABB 02] ABBAS-TURKI A., GRUNDER O., BOUYKHEK R., ELMOUDNI A., "Modular controlled stochastic Petri Nets for the connection monitoring", *Proceedings of the World Automation Congress, Fourth International Symposium on Intelligent Automation and Control*, Florida, 2002.

[ACT 03] ACTIF, *Guide méthodologique – Description d'une architecture multi-projets*, http://its-actif.org, 2003.

[ADA 00] ADAMS BOXILL S., YU L., An evaluation of traffic simulation models for supporting ITS development, Technical Report 167602, Center for Transportation Training and Research, Texas Southern University, 2000.

[AKÇ 01] AKÇELIK R., BESLEY M., "Microsimulation and analytical methods for modelling urban traffic", *Conference on Advance Modeling Techniques and Quality of Service in Highway Capacity Analysis*, California, July 2001.

[ARA 04] ARAMPATZIS G., KIRANOUDIS C.T., SCALOUBACAS P., ASSIMACOPOULOS D., "A GIS-based decision support system for planning urban transportation policies", *European Journal of Operational Research*, vol. 152 (2), 465-475, 16 January 2004.

[ART 04] ARTIMY M.M., ROBERTSON W., PHILLIPS W.J., "Vehicle traffic microsimulator for ad hoc networks research", *International Workshop on Wireless Ad-Hoc Networks (IWWAN)*, 2004.

[ART 07] ARTIMY M., "Local density estimation and dynamic transmission-range assignment in vehicular ad hoc networks", *IEEE Transactions on Intelligent Transportation Systems*, vol. 8-3, 400–412, 2007.

[AWA 04] AWASTHI A., Développement d'un système de routage hiérarchique pour les réseaux urbains, PhD Thesis, University of Metz, 2004.

[BAL 05] BALBO F., PINSON S., "Dynamic modeling of a disturbance in a multi-agent system for traffic regulation", *Decision Support System*, vol.41 (1), 131-146, 2005.

[BAL 07] BALBO F., PINSON S., "A transportation decision support system in agent-based environment", *Journal of Intelligent Decision Technologies*, 1 (3), 97-115, 2007.

[BAL 08] BALBO F., PINSON S. "An agent oriented approach to transportation regulation support systems", *Proceedings 5th Work on Agents in Traffic and Transportation, Autonomous Agents & Multiagent Systems*, 2008.

[BAR 04] BARCELÓ J., CODINA E., CASAS J., FERRER J.L., GARCIA D., "Microscopic traffic simulation: A tool for the design, analysis and evaluation of intelligent transport systems", *Journal of Intelligent and Robotic Systems 41*, 173–203, 2004.

[BAR 06] BARCELÓ J., CASAS J., GARCÍA D., PERARNAU J., "A methodological approach combining macro, meso and micro models for transportation analysis", *Proceedings of the 13th World Conference on ITS*, London, 2006.

[BAR 03] BARREDO J.I., KASANKO M., MCCORMICK N., LAVALLE C., "Modelling dynamic spatial processes: simulation of urban future scenarios through cellular automata", *Landscape and Urban Planning 64*, 145–160, 2003.

[BAT 05] BATTY M., *Cities and Complexity: Understanding Cities with Cellular Automata, Agent-Based Models, and Fractals*, MIT Press, October 2005.

[BEN 04] BENENSON I., TORRENS P.M., "Geosimulation: object-based modeling of urban phenomena", *Computers, Environment and Urban Systems*, vol. 28, nos. 1-2, January–March 2004.

[BEN 06] BENENSON R., PETTI S., FRAICHARD T., PARENT M., "Toward urban driverless vehicles", *International Journal of Vehicle Autonomous Systems*, Publisher Inderscience Enterprises Ltd., 2006.

[BER 02] BERTINI R.L., LINDGREN R., TANTIYANUGULCHAI S., Application of PARAMICS simulation at a diamond interchange, Research Report PSU-CE-TRG-02-02, Portland State University Transportation Research Group, 30 April, 2002.

[BIE 02] BIELLI M., CARAMIA M., CAROTENUTO P., "Genetic algorithms in bus network optimization", *Transportation Research Part C: Emerging technologies*, vol. 10 (1), 19-34, February 2002.

[BLO 00] BLOOMBERG L., DALE J., "Comparison of VISSIM and CORSIM traffic simulation models on a congested network", *Transportation Research Record: Journal of the Transportation Research Board*, vol. 1727, 52-60, 2000.

[BON 01] BONNEL P., Prévision de la demande de transport, HDR thesis, University of Lyon 2, 2001.

[BOU 06] BOUAMRANE K., TAHON C., "Régulation d'un réseau de transport urbain: formalisation des actions de régulation et raisonnement à partir de cas" *6ème conférence francophone de modélisation et simulation, MOSIM*, Rabat, 2006

[BOU 07] BOUSSOUGOU N.S., Evaluation de l'efficacité des transports en commun urbains en site propre, élaboration d'un modèle mathématique d'optimisation du débit horaire, PhD Thesis, University of Lille, 2007.

[BOU 08] BOUYEKHF R., ABBAS-TURKI A.J., GRUNDER O., EL MOUDNI A., "Modelling, performance evaluation and planning of public transport systems using generalized stochastic Petri nets", *Transportation Reviews*, vol. 23 (1), 51-69, 2008.

[BOX 00] BOXILL A., YU L., An evaluation of traffic simulation models for supporting ITS development, Texas Southern University Report 167602-1, Center for Transportation Training and Research, October, 2000.

[CAS 06] CASTLE C.J.E., CROOKS A.T., Principles and concepts of agent-based modelling for developing geospatial simulations, Working Paper 110, Centre for Advanced Spatial Analysis, University College London, 2006.

[CHA 06] CHABROL M., SARRAMIA D., TCHERNEV N., "Urban traffic systems modelling methodology", *Int. J. Production Economics 99*, 156–176, 2006.

[CHA 95] CHAKROBORTY P., DEB K., SUBRAHMANYAM P., "Optimal scheduling of urban transit systems using genetic algorithms", *Journal of Transportation Engineering*, 121(6), 1995.

[CHA 04] CHANARD T., PARENT M., "Quelle place pour la voiture en ville", *La lettre de la prospective*, Sénat, 2004.

[CHI 02] CHIHAIB-BOUZBOUZ F., Approche floue pour la régulation multimodale dans les réseaux de transport urbain en mode perturbé, PhD Thesis, Lille University of Science and Technology, December 2002.

[CHU 08] CHUN W., GE C., YANYAN L., HORNE M., "Virtual-reality based integrated traffic simulation for urban planning", *International Conference on Computer Science and Software Engineering*, vol. 2, 1137-1140, 2008.

[CRA 09] CRAINIC T.G., RICCIARDI N., STORCHI G., Models for evaluating and planning city logistics transportation systems, Report CIRRELT-2009-11, March 2009.

[CYB 04] Expérimentation de cybercars en milieu urbain, Cybernetics Transportation Systems for the City of Tomorrow-WP5, Report, Antibes, 2004.

[DAO 06] DAOUD R.M., EL-DAKROURY M.A., SALLEZ Y., "Wireless vehicle communication for traffic control in urban areas", *32nd Annual Conference of the IEEE Industrial Electronics Society*, IEEE IECON, Paris, November 2006.

[DAO 08] DAOUD R.M., Ethernet câblé et sans-fil pour les systèmes de transports intelligents, PhD Thesis, University of Valenciennes, December 2008.

[DAS] DASH navigation system, http://www.dash.net.

[DEM 08] DEMONGODIN I., HZAMI Z., "Modélisation et analyse de la circulation de véhicules par réseaux de Petri Lots", *Conférence internationale de modélisation et simulation MOSIM*, Paris, 2008.

[DEY 05] DEYMIER G., NICOLAS J.P., Modèles d'interaction entre transport et urbanisme: état de l'art et choix du modèle pour le projet SIMBA, Intermediary Report no. 1, 2005.

[DIF 04] DI FEBBRARO A., SACCO N., "On modelling urban transportation networks via hybrid Petri nets", *Control Engineering Practice 12*, 1225–1239, 2004.

[DON 08] DONIEC A., MANDIAU R., PIECHOWIAK S., ESPIÉ S., "Anticipation based on constraint processing in a multi-agent context", *Journal of Autonomous Agents and Multi-Agent Systems (JAAMAS)*, 17, 339-361, 2008.

[ESS 97] ESSER J., SCHRECKENBERG M., "Microscopic simulation of urban traffic based on cellular automata", *International Journal of Modern Physics C*, vol. 8 (5), 1025-1036, World Scientific Publishing Company, 1997.

[EUR 02] European Commission, Final Report for Publication of the SMARTEST, Simulation modelling applied to road transport European Scheme Test, http://www.its.leeds.ac.uk/projects/smartest/index.html, 2002.

[EZZ 08] EZZEDINE H., BONTE T., KOLSKI C., TAHON C., "Integration of traffic management and traveller information systems: basic principles and case study in intermodal transport system management", *International Journal of Computers, Communications & Control (IJCCC)*, 3, 281-294, 2008.

[FAY 03] FAYECH B., Régulation d'un réseau de transport multimodal: systèmes multi-agent et algorithmes évolutionnistes, PhD Thesis, University of Lille, 2003.

[FOU 02] FOURAGE J.M., Recensement des pannes pour le tramway, Responsable exploitabilité tramway, Semurval Report, 2002.

[FUK 02] FUKUDA T., TAKEFUJI K., IKEMOTO Y., HASEGAWA Y., "Dynamical route-planning for vehicles based on global traffic information and communication", *The IEEE 5th International Conference on Intelligent Transportation Systems*, 538-543, 2002.

[GAR 05] GARAIX T., JOSSELIN D., FEILLET D., ARTIGUES C., CASTEX E., "Transport à la demande points à points en zone peu dense. Proposition d'une méthode d'optimisation de tournées", *SAGEO*, 2005.

[GER 08] GEROLIMINIS N., KUWAHARA M., "The use of telematics to monitor traffic in urban areas: theory and applications", *Intelligent Transportation Systems, 11th International IEEE Conference*, 12-15, Beijing, October 2008.

[GER 06] GERON A., *WIFI, déploiement et sécurité*, Dunod, 2006.

[GIB 09] GIBAUD A., THOMIN P., "FORESEE: a simulation platform for experiments on road trafic self-organization mechanisms", *39th International Conference on Computers & Industrial Engineering*, Troyes, France, 2009.

[GRO 06] GROSS T., NAUMOV V., BAUMANN R., "An evaluation of inter-vehicle ad hoc networks based on realistic vehicular traces", *MobiHoc*, Florence, 22-25 May 2006.

[GUA 98] GUALTIERI G., TARTAGLIA M., "Predicting urban traffic air pollution: a GIS framework", *Transportation Research Part D: Transport and Environment*, vol. 3 (5), 329-336, September 1998.

[HEN 04] HENSHER A., BUTTON K.J., HAYNES K.E., *Handbook of Transport Geography and Spatial System*, Emerald Group Publishing, 2004.

[HOU 06] HOUSSIN L., Contribution à la commande des systèmes Max-Plus linéaires, Applications aux réseaux de transport, PhD Thesis, University of Angers, December, 2006.

[HUI 02] HUISMAN D., FRELING R., WAGELSMAN A., "A robust solution approach for the dynamic vehicle scheduling problem", *Transportation Science*, 22 August 2002.

[ISS 05] ISSARNY V., SACCHETTI D., TARTANOGLU F., "Developing ambient intelligence aystems: a solution based on web services", *Automated Software Eng.*, vol. 12 (1), 101–137, 2005.

[JAY 03] JAYAKRISHNAN R., CORTÉS C.E., PAGÈS L., LAVANYA R., "Simulation of urban transportation networks with nultiple vehicle classes and services: classifications, functional requirements and general-purpose modeling schemes", *Transportation Research Board Annual Meeting*, 2003.

[KAA 07] KAAKAI F., Modélisation et évaluation des pôles d'échanges multimodaux, PhD Thesis, University of Franche-Comté and the University of Technology of Belfort-Montbéliard, December. 2007.

[KAC 02] KACEM L., HAMMADI S., BORNE P., "Pareto-optimality approach for flexible job-dhop scheduling problems: hybridization of genetic algorithms with fuzzy logic", *Journal of Mathematics and Computer in Simulation*, Elsevier Science, 245-276, 2002.

[KHA 07] KHALED F., ABDELGHANY H., HANI S., "Dynamic trip assignment-simulation model for intermodal transportation network", *Transportation Research Record: Journal of the Transportation Research Board*, vol. 1771, 52 – 60, 2001.

[KHA 08] KHALESIAN M., PAHLAVANI P., DELAVAR M.R., "Gis-based multi-agent traffic micro simulation for modelling the local air pollution", *The International Archives of the Photogrammetry, Remote Sensing and Spatial Information Sciences*, vol. XXXVII, part B8, Beijing, 2008.

[KIT 05] KITAMURA R., KUWAHARA M., "Simulation approaches in transportation analysis: recent advances and challenges", *Operations research-computer science interfaces*, vol. 31, 2005.

[LAB 08] LABBOUZ S., Le choix du tracé d'une ligne de transport en commun en site propre et de la position de sa plateforme en milieu urbain, PhD Thesis, University Paris-Est, September 2008.

[LAÏ 02] LAÏCHOUR H., Modélisation multi-agent et aide à la décision: application à la régulation des correspondances dans les réseaux de transport urbain, PhD Thesis on Automatique et informatique industrielle, University of Science and Technology of Lille, 2002.

[LEB 05] LEBACQUE J.P., "Simulation et modélisation des transports", *Forum simulation et transports*, Paris, 2005.

[LEE 04] LEE D.H., BOYCE D., *Urban and Regional Transportation Modeling*, Edward Elgar, 2004.

[LI 91] LI Y., ROUSSEAU J.M., GENDREAU M., Real-time scheduling on a transit bus route: a 0-1 stochastic programming model, Research Report PUB-772, Centre de Recherche sur le Transport (CRT), University of Montreal, May 1991.

[LÓP 08] LÓPEZ-NERI E., RAMÍREZ-TREVIÑO A., LÓPEZ-MELLADO E., "Hierarchical modeling of urban traffic systems for multi-agent based simulation", *Proceedings of the World Congress on Engineering and Computer Science*, San Francisco, 22–24 October 2008.

[LUZ 10] LUZEAUX D., RUAULT J.R., *Systems of Systems*, ISTE, London, John Wiley & Sons, New York, 2010.

[MAS 05] MASAHI S., JUN T. TAKAAKI U., TERUO H., "Evaluation of inter-vehicle ad hoc communication protocol", *Proceedings of the 19th International Conference on Advanced Information Networking and Applications*, vol. 1, 78–83, 2005.

[MED 08] MEDINA J.S., MORENO J.G., ROYO E.R., "Evolutionary computation applied to urban traffic optimization", *Advances in Evolutionary Algorithms*, KOSINSKI W. (ed.), 468, November 2008.

[MEI 07] MEIGNAN D., SIMONIN O., KOUKAM A., "Simulation and evaluation of urban bus-networks using a multiagent approach", *Simulation Modelling Practice and Theory*, vol. 15 (6), 659-671, July 2007.

[MEL 08] MELKI A., Système d'aide à la régulation et évaluation des transports multimodaux intégrant les cybercars, PhD Thesis, École Centrale de Lille, 2008.

[MIL 03] MILLER E.J., HUNT J.D., ABRAHAM J.E., SALVINI P.E., "Microsimulating urban system", *Computers, Environment and Urban Systems*, vol. 28 (1-2), 9-44, January–March, 2003.

[MOB 05] Projet MobiVIP, Véhicules individuels publics pour la mobilité en centre ville, Report, Projet Predit 3 – Intégration des Systèmes d'Information et de Communication, June, 2005.

[MOH 98] MOHAN S.B., TOM V.M., "Urban bus transit route network design using genetic algorithm", *Journal of Transportation Engineering*, 124(4), 368-375, 1998.

[NAD 04] NADEEM T., DASHTINEZHAD S., LIAO C., IFTODE L., 'TrafficView: A Scalable Traffic Monitoring System", *IEEE International Conference on Mobile Data Management (MDM'04)*, 2004.

[NAI 02] NAIT SIDI A., MANIER M.A., EL MOUDNI A., "A controlled policy for public transport network modelled by Petri Nets and Max-Plus Algebra", *Proceedings of World Automation Congress*, Orlando, Florida, 9-13 June 2002.

[NGU 07] NGUYEN DUC K., Contribution à l'étude des problèmes de réordonnancement en cas de perturbations dans un système de transports urbain, PhD Thesis, INPG, October 2007.

[OUL 06] OULD SIDI M., Contribution à l'amélioration des systèmes d'aide à la décision pour la régulation du trafic des réseaux de transport collectif, PhD Thesis, École Centrale de Lille, Université des Sciences et Technologies de Lille, 2006.

[PAR 05] PARK S.J., Bus network scheduling with genetic algorithms and simulation, MSc Thesis, University of Maryland, 2005.

[PAS 02] PASQUIER L., Modélisation des raisonnements tenus en contexte. Application à la gestion d'incidents sur une ligne de métro, PhD Thesis, University Pierre and Marie Curie, Paris, 2002.

[PID 96] PIDD M., DE SILVA F.N., EGLESE R.W., "A simulation model for emergency evacuation", *European Journal of Operational Research*, vol. 90 (3), 413-419, 10 May 1996.

[PUR 01] PURSULA M., NIITTYMÄKI J., *Mathematical Methods on Optimization in Transportation Systems*, Springer, 2001.

[RAD 03] RADWAN E.R., MOLLAGHASEMI M., Framework For Modeling Emergency Evacuation, Project ID #58, Center for Advanced Transportation Systems Simulation, University of Central Florida. Orlando, 2003.

[RIE 03] RIES R., Transports urbains: quelles politiques pour demain, Report, July, 2003.

[ROU 03] ROUSSELOT M., "Transports et développement durable: vers des politiques innovantes en milieu urbain", *Annales des Mines*, 2003.

[RUB 07] RUBIO L., REIGA J., RODRIGO-PEÑARROCHAA V.M., CARDONAA N., "A semi-deterministic propagation model for predicting short-term fading statistics in urban environments based on the Nakagami-*m* distribution", *AEU-International Journal of Electronics and Communications*, vol. 61 (9), 595-604, 1 October 2007.

[SAI 04] SAITO M., FUNAI M., UMEZU T., Inter-vehicle mobile ad-hoc communication protocol for acquiring local traffic information, IEIC Technical Report, Institute of Electronics, Information and Communication Engineers, vol. 103, 49–56, 2004.

[SAL 97] SALIM V., CAI X., "A genetic algorithm for railway scheduling with environmental considerations", *Environmental Modelling and Software*, vol.12 (4), 301-309, 1997.

[SAR 07] SART, Système d'aide à la régulation de réseaux de transports multimodaux, Report, December 2007.

[SCH 07] SCHMIDT-EISENLOHR F., TORRENT-MORENO M., MITTAG J., HARTENSTEIN H., "Simulation platform for inter-vehicle communications and analysis of periodic information exchange", *IEEE Fourth Annual Conference on Wireless on Demand Network Systems and Services*, 50–58, 2007.

[SCH 08] SHIE-YUAN W., CHIH-CHE L., "NCTUns 5.0: A Network Simulator for IEEE 802.11(p) and 1609 Wireless Vehicular Network Researches", *69th Vehicular Technology Conference*, 2008.

[SHI 99] SHIOSE T., SAWARAGI T., KATAI O., "Multiagent model of VICS", *IEEE International Conference on Systems, Man, and Cybernetics, IEEE SMC'99 Conference Proceedings*, vol. 4, 574–579, 1999.

[SKY 08] Projet Skyweb Express, http://www.taxi2000.com, 2008.

[TAY 01] Taylor M., "Intelligent Transports Systems", *Handbook of Transport Systems and Traffic Control*, BUTTON K.J., HENSHER D.A. (eds), Pergamon, Oxford, 2001.

[THÉ 03] THÉVENIN T., FAUVET M.C., JOSSELIN D. "Régulation spatio-temporelle d'un réseau de transport public. Articulation intermodale d'un réseau de bus, d'un transport à la demande et d'un espace piétonnier", *Revue Internationale de Géomatique, SIG et Transport*, 161-180, Hermes, 2003.

[THO 12] THOMIN P., GIBAUD A., KOUTCHERAWY P., "Deployment of a fully distributed system for improving urban traffic flows: a simulation based performance analysis", *Simulation Modelling Practice and Theory*, vol. 31, pp. 22-38, 2012.

[TIB 07] TIBOR L., Y. ZANG, REUMERMAN H.J., "Neighborhood evaluation of vehicular ad-hoc network using IEEE 802.11p", *13th European Wireless Conference*, Paris, 1-4 April 2007.

[TOI 05] TOILIER F., ALLIGIER L., PATIER D., ROUTHIE J.L., Vers un modèle global de la simulation de la logistique urbaine: FRETURB, version 2, Report, 2005.

[TOL 05] TOLBA C., LEFEBVRE D., THOMAS P., EL MOUDNI A., "Continuous and timed Petri nets for the macroscopic and microscopic traffic flow control", *Simulation Modelling Practice and Theory*, vol. 13, 407-436, 2005.

[TOR 07] TORRENT-MORENO M. "Inter-vehicle communications: assessing information dissemination under safety constraints", *Fourth Annual Conference on Wireless on Demand Network Systems and Services, WONS'07*, vol. 4, 59–64, 2007.

[TRE 00] TREIBER M., HENNECKE A., HELBING D., "Congested traffic states in empirical observations and microscopic simulations", *Physical Review E 62*, 1805-1824, 2000.

[ULT 08] ULTra PRT System, http://www.atsltd.co.uk/, 2008.

[WAD 04] WADDELL P., ULFARSSON G.F., "Introduction to urban simulation: design and development of operational models", *Handbook in Transport, Volume 5: Transport Geography and Spatial Systems*, STOPHER, BUTTON, KINGSLEY, HENSHER (eds.), Pergamon Press, 203-236, 2004.

[WIS 04] WISCHHOF L., EBNER A., ROHLING H., "Self-organizing traffic information system based on car-to-car communication: prototype implementation", *Proceedings of the 1st International Workshop on Intelligent Transportation (WIT2004)*, Hamburg, Germany, 2004.

[WIS 07] WISETJINDAWAT W., SANO K., MATSUMOTO S., RAOTHANACHONKUN P., "Micro-simulation model for modeling freight agents interactions in urban freight movement", *86th Annual Meeting of the Transportation Research Board*, Washington D.C., 21-25 January 2007.

[YAO 06] YAO J., TAWFIK H., FERNANDO T., "A GIS based virtual urban simulation environment", *ICCS2006*, Part III, LNCS 3993, 60–68, 2006.

[YIN 02] YIN H., WONG S.C., XU J., WONG C.K., "Urban traffic flow prediction using a fuzzy neural network approach", *Transportation research Part C.*, Elsevier Science Ltd., 85-98, 2002.

[YU 05] YU B., YAN Z., LIU C., CHENG C., "Optimization of bus transit network with parallel ant colony algorithm", *Proceedings of the Eastern Asia Society for Transportation Studies*, vol. 5, 374-389, 2005.

[ZAN 03] ZANARELLI C., Caractérisation des stratégies instrumentales de gestion d'environnements dynamiques. Analyse de l'activité de régulation du métro, PhD Thesis, University of Paris VIII – Saint Denis, 31 January 2003.

Chapter 3

Real-time Fleet Management: Typology and Methods

3.1. Introduction

Current information and communication technologies provide new data streams in real time, which have been used in recent applications particularly in the field of transport. More specifically, in terms of fleet management, a carrier may offer an improved Quality of Service and manage its business more effectively. It can, for example, know the exact position of each of its vehicles, at any time, and be in contact with them in a bi-directional manner (provide information to drivers, confirm the completion of a task), etc. Processing of this data set leads to better responsiveness to any changes in the environment and the state of the fleet, and better monitoring of traceability of goods, etc. Before continuing, let us clarify the semantics of the terms used here and throughout this chapter. We mean by:

– *request*: a service request performed by the carrier that requires the intervention of at least one vehicle in the fleet for its realization;

– *mission*: any action carried out with a vehicle to achieve one or more requests;

Chapter written by Frédéric SEMET and Gilles GONCALVES.

– *route*: a series of missions carried out or planned for a vehicle on a given time horizon.

The real-time fleet management problem (RTFMP) involves:

– assigning every new request to a route while previously scheduled routes are still in progress;

– ensuring the smooth-running of routes to complete scheduled missions.

We consider the case of dynamic problems for which we have to make decisions based on only partially known information, and modify these decisions when new information is received.

After a general overview regarding the context of dynamic problems for real-time fleet management in section 3.2, we describe two approaches based on metaheuristics for real-time fleet management in sections 3.3 and 3.4. Finally, we draw our conclusions in section 3.5.

3.2. General context of RTFMP

In this section, we briefly present the general characteristics of RTFMP and its application contexts; for more details we refer the reader to [BIA 00; PIL 12].

3.2.1. *RTFMP characteristics*

Real-time fleet management problems fall within the scope of dynamic transport problems. To recall the definition of a dynamic problem, we include one proposed by Powell, Jaillet and Odoni [POW 95]:

> "A problem is dynamic if decisions must be made before all the information is known and change as new information is received."

In addition, we can distinguish between two types of problem where data changes over time, but only those of the first type are dynamic problems: problems with *dynamic data*, i.e. those for which the information is constantly changing, such as the set of customers to serve, etc., problems with *time-dependent data*, i.e. those for which the information is known in advance but is dependent on time. For example, the delivered quantity can be different depending on the time at which a customer is served.

To clarify the differences between the real-time fleet management problem and vehicle routing problems, we propose a classification of the characteristics listed by Psaraftis [PSA 95]. We distinguish four groups of features:

1) *State-observation*: this subset consists of features related to the state and the observation of the fleet of vehicles. It includes: i) the *temporal dimension*, meaning the possibility for the carrier to know at any time the vehicle's position and the nature of the activity being performed; ii) *update of information*; iii) *size of the fleet*: the number of vehicles used in a RTFMP is usually fixed and has a direct impact on activity, in contrast to classical transport problems where the fleet size is a decision variable.

2) *Prediction-anticipation:* this RTFMP characteristic is related to future knowledge or management. It is related to the *time horizon* and the *nature of future information*. Several *time horizons* can be considered for the management of the fleet: short horizons where only known or expected requests, to be addressed in the near future or even only the next request (as in the case of emergency services), are considered; rolling horizons where the requests, known or predicted for a time fixed period, are taken into account; long horizons where all known or anticipated requests are considered. The future can be modeled in two distinct ways. In non-stochastic RTFMPs, only known factors are taken into consideration, while in stochastic RTFMPs, probability distributions are known about the events that may occur in the future, such as the arrival of a new customer. These distributions are, for example, obtained by using statistical data from the management of the history of past events.

3) *Evaluation*: based on system state information predictions, the carrier assesses the situation and possible routing changes in order to make certain decisions. It is characterized by: *the short-term management*, *the computation time available* and *the objective function*.

4) *Decision*: this final group makes decisions following the evaluation. It is characterized by: the assignment of a task or sequence of tasks, the acceptance delay, the queueing management and the nature of time constraints.

3.2.2. *Application field of RTFMPs*

The scope of application of real-time fleet management problems is quite large. As an overview, we use the method proposed by Gendreau and Potvin [GEN 06], summarized in Table 3.1.

Consolidation \ Territory	Local	National
Yes	Letter and parcel delivery service Transport of elderly or disabled persons	Less-than-truckload trucking
No	Emergency services (taxi, ambulance, fire engine, police, tow truck, etc.)	Full load trucking

Table 3.1. *Classification of dynamic fleet management problems*

For problems without a *consolidation* component, a repositioning problem within the fleet can be considered to anticipate the future requests. This is particularly the case for the ambulance fleet management problem [GEN 01], which consists of redeploying available ambulances after assigning one of them to respond in less than 15 minutes to any request, and in less than 7 minutes for 90% of the most urgent cases.

Truck fleet management problems involve moving goods between two locations, while respecting time and capacity constraints. Trucking problems can be divided into two categories:

– only a single request can be processed at a time, i.e. any task necessary to achieve an application must be completed before starting another request. We then talk of a *full load trucking* problem [YAN 04];

– multiple requests can be processed simultaneously, i.e. missions that are required to achieve several requests can be integrated. We then talk of a *less-than-truckload trucking* problem [MAL 04].

For each of these problems, a request may be rejected if some constraints cannot be met, or if its implementation is too expensive for the carrier, which constitutes a significant difference to the management of the ambulance fleet. In terms of objectives, a hierarchy of multiple objectives such as maximizing the number of accepted requests and minimizing the duration, distance, total cost of route, or the total delay is frequently defined. The proposed methods are based on: modifications of assignments of accepted requests in order to meet a new request, and less-than-truckload trucking problems on modifications in the sequences of tasks to perform.

The last class of problems is that of letter and parcel delivery problems [GEN 06; MIT 04]. These consist of a vehicle collecting letters or parcels and reporting to a sorting center, or delivering after sorting. This is a less-than-truckload trucking problem, where the geographical area to be covered is local, and capacity constraints are not present. The problem of transporting elderly or disabled persons differs from the previous problem due to the importance of complying with time slots selected by customers who are transported from door to door, and who possibly make their request shortly before the time when it should be performed. An objective of the problem is then to minimize the time between the start of a request and its realization [COR 03].

3.3. Simulation platform for real-time fleet management

In this section, we describe an experimental platform that we have developed as part of our academic research on the real-time management of fleet vehicles. This work took place over several phases:

– definition of a software architecture to optimize and simulate the course of real-time fleet management;

– consideration of the uncertainties on the quantities to be delivered to customers;

– consideration of traffic information to estimate travel time.

The problem we chose to process in these works addresses the management of a Vehicle Routing Problem with Time Windows (VRPTW), but this work has been stretched to other extensions of the vehicle routing problem in H. Housroum's thesis [HOU 04].

To experimentally test our solution approaches, we developed a Java simulation platform with the ability to simulate a day's service, taking into account both the dynamic aspects (i.e. arrival of a new request), the uncertainty of systems (i.e. imprecise delivery quantity) and the variability of measured phenomena (i.e. fluctuating travel time).

3.3.1. *Dynamic management of vehicle routing*

In a dynamic vehicle routing problem, we generally distinguish the so-called "static" customers, meaning customers known in the preliminary scheduling phase, and so-called "dynamic" customers, meaning those who will only be known during the live processing phase (i.e. execution of routes).

In Figure 3.1, we have a simple example of a dynamic problem where two vehicles must serve a set of customers. Static customers are represented by black nodes, while dynamic customers (i.e. new customers) are represented by white nodes. Thin line arcs represent the scheduled route for each vehicle. The two thick solid line arcs indicate current vehicle trips from when the dynamic requests were received. New customers must be inserted into the existing routes, taking into account yet unvisited customers (dashed line).

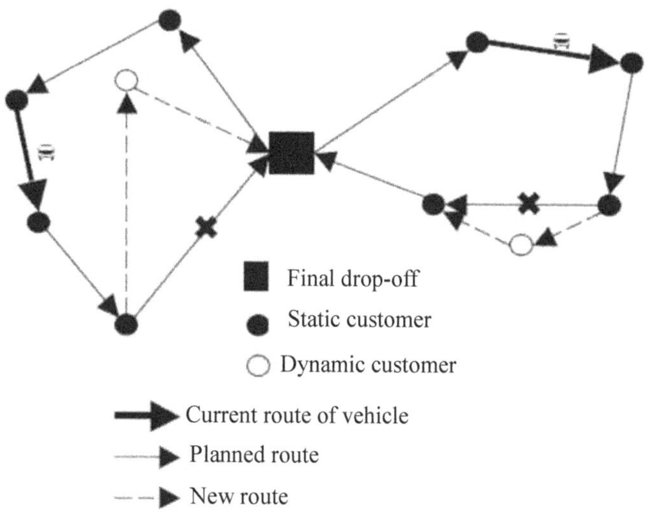

Figure 3.1. *A dynamic vehicle routing problem*

Unlike a static vehicle routing problem, the performance of the dynamic management used not only depends on the number of customers and the spatial distribution of them, but also on the number of dynamic events and times when these events take place. Therefore, a measure to describe the dynamics of the system is very useful when we want to examine the performance of a specific algorithm under varying conditions. The notion of a degree of dynamism (dod) has been defined as below [LUN 96]:

$$dod = \frac{Number\ of\ dynamic\ customers}{Total\ number\ of\ customers}$$

However, this measure does not take account of the arrival times of immediate requests (i.e. dynamic customers) that arrive during the day. This means that two different systems will be encountered: one in which immediate requests are received at the beginning of the planning horizon and one in which they appear throughout the day.

To overcome this problem, a second indicator was introduced in the work of Larsen [LAR 00]. The number of immediate requests received during the planning horizon is denoted n_{imm} and the number of static requests is denoted n_{adv}. The total number of requests n_{tot} is equal to $n_{imm} + n_{adv}$. The effective degree of dynamism, denoted by *edod*, is then defined by the following expression:

$$edod = \frac{\sum_{i=1}^{n_{imm}} \left(\frac{tr_i}{T}\right)}{n_{tot}}$$

The effective degree of dynamism represents the average delay of each request received in relation to the maximum allowable delay defined by the planning horizon T.

3.3.2. *Routing management under time window constraints*

The Vehicle Routing Problem with Time Windows (VRPTW) is a variant of the VRP, where an additional constraint (i.e. time window) is added to the start time of service to the customer. Each customer Ci is allocated a time window [ei, li] which represents a time interval in which a customer requests to be served. It should be noted that all these problems (VRP and VRPTW) are NP-hard and their complexity has been demonstrated by Lenstra and Rinnooy Kan [LEN 81]. Similarly, when the fleet size is fixed, the problem of the existence of a feasible solution is a NP-complete problem. This time constraint greatly complicates the resolution of the problem and only relatively small instances can be solved precisely. The most commonly used methods are those of branch and cut, column generation or Lagrangian relaxation. For larger instances, researchers initially focused on heuristic methods but now focus on metaheuristics (tabu search, genetic algorithms, simulated annealing and ant colonies) which can provide approximate solutions of acceptable quality in reasonable computation times.

In summary, the time window constraint further complicates the vehicle routing management problem, and finding a solution may require the commitment of additional vehicles that are not necessarily available. Compromises must be found between customer satisfaction (i.e. percentage of customers served) and distribution costs. Also, some authors propose relaxing the time window constraint, accepting, through penalties, solutions where the customer would be served outside the time window. These scientific studies have therefore given rise to two variants of the problem [TAI 97]: the VRPTW with hard time window constraints (VRPHTW – Vehicle Routing Problem with Hard Time Windows) and soft constraints on VRPTW with time windows (VRPSTW – Vehicle Routing Problem with Soft Time Windows).

In our work, we are interested in the vehicle routing problem with soft time windows because it provides better applicability to business problems when the fleet size is limited.

3.3.3. *General architecture of the simulation platform*

The platform we developed [GON 07] was written in Java language because it enables rapid development of prototypes. It relies on two main components that work together (Figure 3.2): a simulator component that simulates the real world, generating significant daily events, and a planning component that is responsible for instantly providing an appropriate response to external events. The latter includes three components: a resolution method component that applies the selected resolution strategy, a component that estimates travel time, and an event management (EM) component that waits for an event and triggers appropriate actions for the event. A graphic interface allows us to visualize the evolution of routing in the planning horizon. In our case, the resolution method component is based on a genetic algorithm (GA), but the architecture of the platform can be applied to other solution strategies and therefore compare their respective performances. To summarize, our tool is a platform that works “live” (i.e. awaiting new events) and the resolution module constantly searches, without user intervention, to improve existing routes (i.e. in a background way).

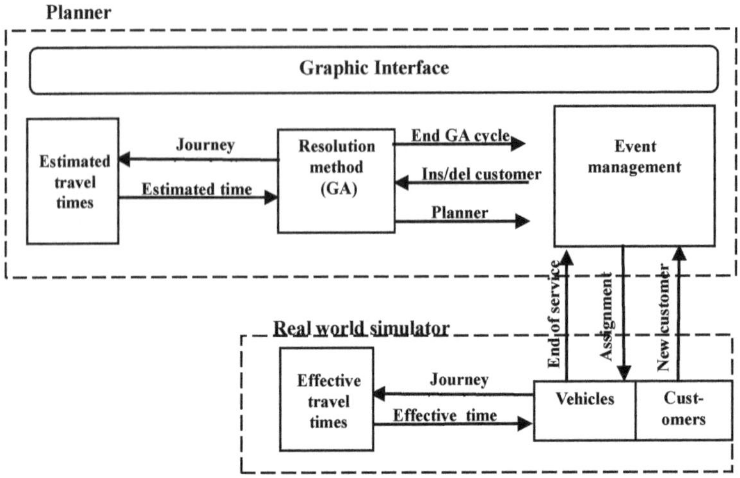

Figure 3.2. *Java platform simulation architecture*

To dynamically determine vehicle routes, we favored a metaheuristic based on GAs that can quickly find appropriate solutions to the problem. GAs are approximate resolution methods that explore the search space by neighborhoods using a population of individuals on which we apply selection, mutation and crossover operators to adapt them to their environment [GOL 89]. GAs have been previously used to solve the VRP [PRI 04] and static VRPTW [THA 95] with very good performance. Moreover, the adaptive nature of genetic algorithms provides good processing of a difficult optimization problem in an evolving context where data changes over time (i.e. number of customers, travel time, etc.) [BRA 01]. At each modification, the optimal solution to the initial problem may or may not be affected by this disruption. If it fails, an adaptation of the former solution is needed. The classical approach to dealing with this dynamic problem is to consider that each change creates a new optimization problem that must be solved from scratch. However, although this approach is simple, it is often not applicable in a context of tight time constraints. In addition, we observed that the solution of a new problem is not too different from the solution of the old problem when the environment is only slightly disturbed, which is generally the case in our application. That is why we chose to update current solutions to provide new solutions that take into account the encountered disturbance.

In the basic version of our platform, two types of event are taken into account (Figure 3.2): the arrival of a new customer request and the end of service with another customer. The "new customer" event uses an insertion heuristic to add the new customer to the best position of each chromosome of current population. As the number of customers to serve is unknown, the size of the chromosome is variable. At any time, the chromosome shows the number of customers remaining to be served in the future. Note that the sequence of customers in the corresponding genotype defines an order of preference in the allocation of customers to available vehicles. The encoding used by our GA is indirect and to obtain the corresponding solution for a chromosome, an insertion heuristic is used. To handle the customer "end of service" event, the planner must provide the corresponding vehicle with the next customer to serve ("assignment" answer in Figure 3.2). The individual in the population with the highest "fitness" (i.e. providing the best solution) is chosen. If there are no more customers to be served by this vehicle, it returns to the depot.

To evaluate this resolution approach, we used instances of problems for 100 customers proposed in Solomon's benchmark [SOL 87]. This constitutes a baseline for calibrating the performance of most methods proposed for treating VRPTW. These instances have been adapted by Gendreau *et al.* [GEN 99] to add the dynamic nature of the arrival of customer requests. The objective function used is a compromise between the total distance covered by the fleet of vehicles and the sum of customer delays (i.e. customers served outside their time windows).

We compared our results using a simple insertion heuristic and a parallel tabu method, as used in the work of Gendreau *et al.* [GEN 99]. It should be noted that for the VRPTW problem, the tabu method provides the best solutions to static problems. The comparison was made on three criteria (see Table 3.2): the average distance (D), the average delay (R), and the average number of customers who are not served (C). Each value represents the average of the resolutions of instances belonging to the same group of problems (R1, C1, RC1, R2, C2 and RC2). The R group corresponds to a uniform distribution of customers in space while the C group corresponds to a distribution of customers in disjoint clusters. The RC group is a combination of the two previous distribution modes. The issue focuses on the size of the time window for each customer. Type 1 implements narrow time windows over a short service horizon with routes containing few customers. Vehicle capacity is also reduced. Type 2 implements large time windows. As the horizon is widest, routes normally include more customers and vehicle capacity is large.

Problem		Insertion	Tabu search	DGA
C1 Dynamic	D	1234.24	857.20	**840.70**
	R	365.44	1.30	1.44
	C	8.40	0.00	0.00
R1 Dynamic	D	1297.76	1210.80	**1193.10**
	R	393.88	51.70	**51.07**
	C	10.70	0.58	0.75
RC1 Dynamic	D	1393.27	1360.30	**1324.74**
	R	573.19	65.90	80.02
	C	8.40	0.50	**0.03**
C2 Dynamic	D	798.81	628.70	**618.63**
	R	282.08	0.00	0.89
	C	0.13	0.00	0.00
R2 Dynamic	D	1199.84	1050.40	**1047.22**
	R	523.01	40.40	**18.62**
	C	0.51	0.45	**0.00**
RC2 Dynamic	D	1388.48	1192.60	**1192.53**
	R	522.93	43.40	**25.71**
	C	0.58	0.00	0.00
All	D	1218.73	1050.00	**1036.15**
	R	443.42	33.78	**29.62**
	C	4.78	0.26	**0.13**

Table 3.2. *The DVRPTW problem: comparison between different methods*

This comparison shows the good performance of our GA dynamic approach when faced with a greedy-type approach (insertion heuristic). The quality of solutions is comparable or better those (in some of the solutions) provided by the parallel tabu method by Gendreau *et al.* [GEN 99], using a single processor.

3.3.4. *Consideration of uncertainties on requests*

Previous work has been extended to take into account uncertainties found in some vehicle routing applications. Much research has been devoted to various aspects of the VRP but most often in a deterministic context which assumes that the problem data are accurately and completely known. However, in many practical situations, the data from the VRP are not fully known at the time of planning. The Stochastic Vehicle Routing Problem (SVRP) has interested both researchers and industrialists for about 10 years. The VRP is called stochastic when some parameters of the problem are random. This could be due to customer requests (i.e. quantities to deliver or pick up), travel times, cost of transport or the set of customers to visit. The VRP with stochastic demands (VRPSD) is the most studied case. Customer requests are not known in advance, that is to say at the time of planning. Two approaches to model these uncertain requests are possible. In the first approach, requests follow a probability distribution known as the normal distribution in most studies [TAN 07]. In the second approach, the authors are interested in another variant of the VRP: VRPFD (Vehicle Routing Problem with Fuzzy Demand) applications where uncertainties are modeled using fuzzy numbers [TEO 96; WER 03].

In our work, we were interested in cases where uncertainty refers to the quantity requested by each customer that can be found, for example in the supply of heating oil. Customer requests are then represented by fuzzy numbers, denoted $\tilde{d}_i$ [XU 08].

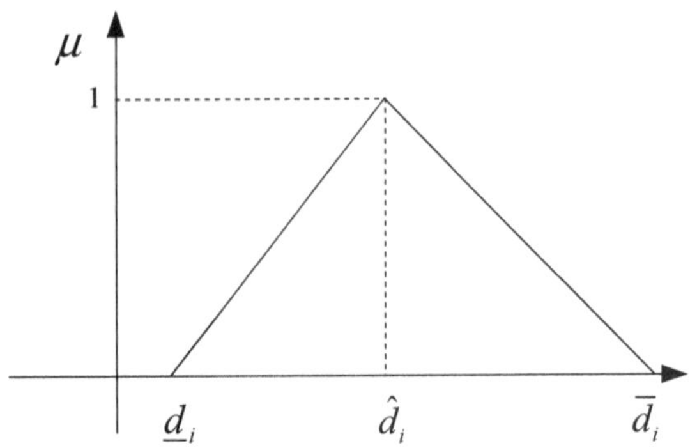

Figure 3.3. *Fuzzy request model*

This fuzzy number is composed of a triplet of real numbers $(\underline{d}_i, \hat{d}_i, \overline{d}_i)$ with $\underline{d}_i < \hat{d}_i < \overline{d}$. A membership function μ is associated with each fuzzy number; it has a triangular representation that reflects a distribution of customer request (see Figure 3.3). The terms $\overline{d}$ and $\underline{d}_i$ are respectively the lower and upper limits, their corresponding membership degrees are equal to 0. The term $\hat{d}_i$ is the normalization corresponding to the maximum degree of membership which is equal to 1. The greater the degree of membership, the more it reflects the confidence of the expert (i.e. by the customer, for example) in the value of the corresponding amount requested.

It is worth noting that in this problem, the actual quantities of customer requests will only be known accurately when vehicles arrive at their customers. Despite this, the vehicle routes should be planned in advance by an algorithm that takes this uncertainty into account. The capacity constraint, which must be respected, becomes a fuzzy capacity constraint. Routes are planned by comparing fuzzy requests with a vehicle of fixed and known capacity. One consequence is that sometimes the vehicle cannot serve a customer on the planned route: for example, when the residual capacity of the vehicle is insufficient to serve the customer. This situation is called a "route failure" [DRO 86]. It happens particularly for customers located later in the route. In Figure 3.4, a route defect appears on one of the three rounds.

The policy action chosen in this case is to return the vehicle to the depot to unload or restock when a lack of capacity occurs with a customer (see road use as dotted line in Figure 3.4). Then the vehicle returns to the customer to complete the route. Other policy actions were defined to overcome the defects routes (preventive restocking, re-optimization of residual tours, etc.) [TAN 07].

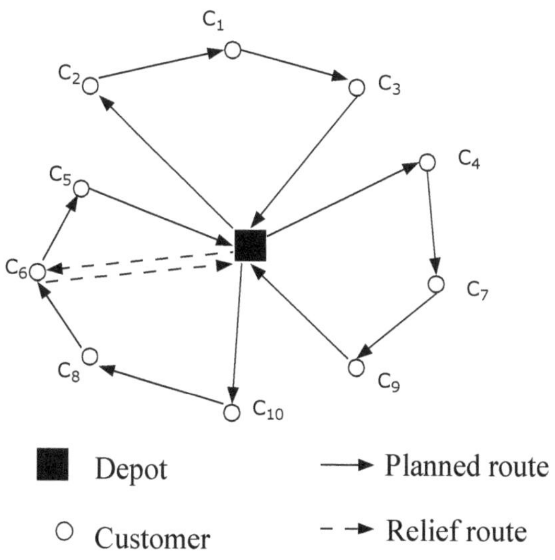

Figure 3.4. *Route failure*

To model this problem, we were inspired by works on the VRPSD (Vehicle Routing Problem with Stochastic Demand) [GEN 96]. A Stochastic Programming with Recourse (SPR) was adapted to take account of fuzzy capacity constraints using the possibility theory introduced by Zadeh [ZAD 78] and developed by Dubois and Prade [DUB 88]. In this model, the VRPSD is solved in two steps: in the first step, a solution is planned based on uncertain data; in the second step, we try to estimate the cost of corrective actions to address "route failure" situations. Optimization involves minimizing the cost estimate provided by the first step, taking into account the additional costs generated by the recourse determined in the second step. The estimated cost can be determined based on a minimum necessity threshold, to be respected for fuzzy capacity constraint. Then the routes produced by the GA are replayed on a set of learning scenarios using a simulator in order to determine the additional cost due to recourse. The best solution (estimated cost + additional cost) is then chosen as the route to be carried out. The learning scenarios are determined by a random selection of customer requests that meet the request distributions selected in modeling applications (i.e. rate of uncertainty of customer request). It should be noted that the necessity threshold, ranging between 0 and 1, represents the confidence that is desired in the planned solution. The higher the number, the more secure the planning will be in order to prevent defect routes, and the more costly the solution will be in terms of resources. A compromise between estimated cost and additional cost can be found by experimental tests on the value of the necessity threshold parameter.

To evaluate the aptitude of our stochastic model with fuzzy requests (denoted by SPR), we compared it with two other models on a set of test scenarios. These were generated as learning scenarios and have been integrated into each of Solomon's benchmark problems, edited by Gendreau as described above. The three models in this comparison use the same GA Java planning platform. A reference model (denoted by Det) assumes that the actual customer requests are known in advance. This model is expected to provide a better performance because it knows customer requests perfectly. A second model (denoted by Nor) is based on the normalized value of the fuzzy customer request. In the latter case, planning is done without taking the uncertainty of a customer request into account (i.e. basic model). This model provides solutions that are not as good.

In Table 3.3, we report the deterioration rate of the solution provided by the stochastic model (t^{SPR}) and the normalized model (t^{Nor}) with the reference model on a subset of representative benchmark problems. We add the prefix S to the original Solomon problems to indicate that this is a static problem where all customers are known at the time of planning.

Type	Problem	t^{SPR}	t^{Nor}	Type	Problem	t^{SPR}	t^{Nor}
SC1	SC101	**2.06%**	**2.06%**	SC2	SC201	**4.81%**	8.49%
	SC105	**2.09%**	6.41%		SC205	**4.50%**	7.70%
	SC108	**1.22%**	10.51%		SC207	5.19%	**3.77%**
	SC109	**4.93%**	5.11%		SC208	**3.64%**	6.85%
	Average	**2.58%**	6.02%		**Average**	**4.54%**	6.70%
SR1	SR101	**1.93%**	10.21%	SR2	SR205	**0.78%**	1.58%
	SR103	**-0.01%**	4.36%		SR206	**0.41%**	0.42%
	SR104	**0.01%**	5.06%		SR207	**1.14%**	2.42%
	SR109	**0.31%**	5.84%		SR208	**2.46%**	9.07%
	Average	**0.56%**	6.37%		**Average**	**1.20%**	3.37%
SRC 1	SRC101	**2.45%**	2.80%	SRC 2	SRC202	**0.90%**	5.69%
	SRC104	**4.26%**	14.45%		SRC204	**2.46%**	3.70%
	SRC106	**1.58%**	6.74%		SRC205	**0.44%**	2.10%
	SRC107	**1.96%**	7.46%		SRC206	**-0.28%**	6.74%
	Average	**2.56%**	7.86%		**Average**	**0.88%**	4.56%
All	**Average**	**2.05%**	5.79%				

Table 3.3. *Comparison of SPR and Nor models with the reference model (for static customers)*

The values shown are averaged values, obtained from five trials, given the non-deterministic nature of genetic algorithms. The value of the uncertainty rate of customer requests is drawn randomly according to a uniform distribution ranging between 0 and 1/3. For all tests, the degradation of the stochastic model with recourse (SPR) compared to the reference model is only 2%. This low value shows the ability of the SPR model to deal with fuzzy requests; the quality of solutions is close to those of the reference model, which is assumed to accurately know the actual customer request. The Nor model does not take into account the uncertainty of a customer and has a degradation of 5.8%, much higher than the SPR model.

The graph in Figure 3.5 shows the evolution of this degradation as a function of the rate of uncertainty of customer request. The more the uncertainty increases, the more the degradation of the model increases and the gap between the SPR and Nor models widens.

We then extended this study to include immediate customers, but the gains are more modest because the uncertainty accumulated on a route is smaller in such a case.

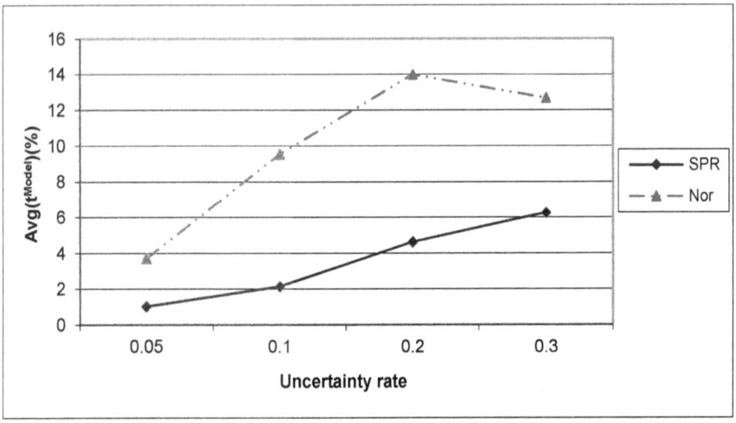

Figure 3.5. *The influence of the rate of uncertainty of request over measured degradation*

3.3.5. *Consideration of information linked to traffic*

Current on-board technologies (GPS) allow real-time tracking of vehicles; it is therefore possible to know the instant speed of vehicles throughout their routes. This information can be uploaded to a Central Station and taken into account by the dynamic route planning system to consider the hazards of vehicular traffic.

In classical vehicle routing problem resolution models, the average velocities of vehicles are generally considered to determine the travel time between two consecutive customers. These are derived from histories. It is assumed that they are generally not subjected to stochastic variations, and velocity is considered constant for the whole planning horizon. In a real situation, the variability of traffic conditions strongly influences the performance of the planned solution. This cannot be ignored if we want to get a good quality route optimization.

Another extension of our work concerns the consideration of real-time traffic conditions and in particular the change in vehicle speed or duration of associated journeys.

To take the variability of travel time into account, time-dependent travel time models have been developed [ICH 03], considering periods of the day called "off-peak" (normal average speed) and "peak" periods (low average speed). A velocity profile in the form of a piecewise constant function is defined for each type of road segment.

The main feature of previous models is that they are based on "static" velocity profiles that do not change based on actual road traffic conditions that are possible to follow due to integrated technologies. To get closer to real problems, it is necessary to extend these models such that the calculated travel time takes into account the emergence of new situations where unforeseen traffic events occur during the day.

Chen [CHE 06] created a dynamic journey time profile that incorporates variations in traffic conditions. However, it should be noted that the travel time profile in this model is not a continuous function, as opposed to the Ichoua *et al.* model [ICH 03]. The First-In-First-Out (FIFO) hypothesis is not met; a vehicle leaving from the same source later than a second vehicle can reach the destination earlier.

Time-dependent travel time models, as described above, consider traffic changes in real time but do not make predictions about the future state of traffic. Vehicle routes are planned and based on current traffic conditions on the assumption that they will remain the same in the future. In many cases, vehicles' routes will be carried out under conditions that do not meet the conditions used in traffic planning.

In our GA Java planning platform, we have defined and integrated a model to calculate travel time, which includes the following three components [ZHA 09]:

– a piecewise constant speed model, depending on the time of day;

– a vehicle monitoring system which receives real-time information on vehicle speed;

– a predictive model designed to provide projections on the future state of traffic.

The change in actual velocities observed in the field is an irregular continuous function which makes it very difficult to obtain precise travel times. To develop our time-dependent real-time model and calculate real travel times, the actual velocity variation is modeled as a piecewise constant function. This function is called the velocity profile; it is shown in the upper left part of Figure 3.6. The planning horizon (i.e. day) is divided into several periods and the velocity is variable throughout the day, but constant within each time period.

In our problem, we assume that changes in actual velocities are uploaded to the planning system in real time thanks to the simulator component of our platform. The equipment on board the fleet vehicles (GPS, GSM, etc.) or sensors integrated on some road sections can contribute by regularly providing information on traffic conditions or the average speed of vehicles observed. That is why it would be useful to use this information to determine a predicted velocity for the subsequent time periods. Consequently, it is necessary to define a model for predicting velocity that we will use in the subsequent time period. For that, we take a simple exponential smoothing model because it is inexpensive in terms of computing time, very responsive to changes in the smoothing coefficient used and therefore fits very well in a real-time environment.

For each period T and for a given route joining two customers, the vehicle velocity is measured and sent back to the planning system in real time. It allows the calculation of a predicted velocity for the next period T+1. This predicted velocity (period T+1) and the history of velocities of the remaining periods (T+2, T+3, etc.) can together form a new velocity profile for this itinerary which will be dynamically updated in each velocity sampling period. These new velocity profiles are then used to re-evaluate, at the same refresh rate, the travel time for each pair of customers to integrate into the current routes (*Origin-Destination* O-D matrix). The O-D matrix is then used by the "estimated travel time" component of our platform (Figure 3.2). The GA component urges the latter to try to improve future routes continuously (i.e. in the background).

Figure 3.6 describes an example for developing a new planning profile. For simplicity, we assume that the day of service is divided into nine time periods. The historical velocity profile is a piecewise constant function consisting of three parts: the first, second and third parts correspond respectively to the periods between T and T+3, T+3 and T+6, and between T+6 and T+9. The number of time periods used to measure and predict the rate of velocity equals nine periods. At time T+1, the speed interval is measured between T and T+1 (vm is the speed), then the speed is predicted for the interval between T+1 and T+2 (vp is the speed). In this way, we obtain the new velocity profile to be used for planning routes. We call this new profile the "planning profile". This is the planning profile that is used by the "estimated travel time" component to estimate travel time.

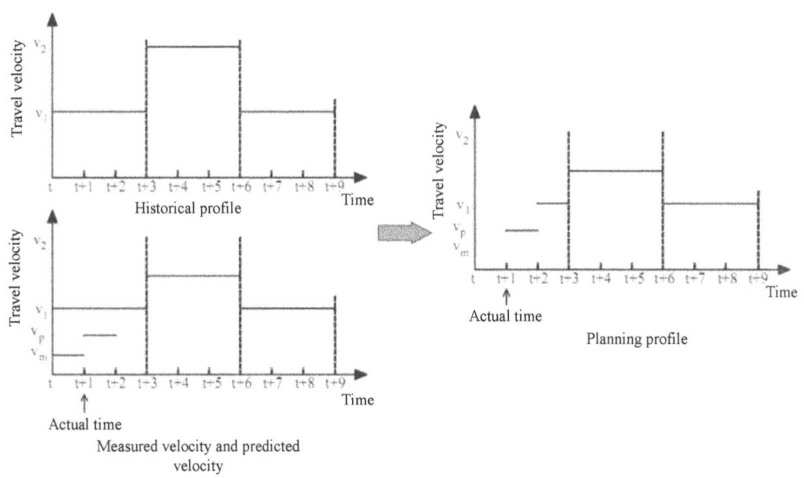

Figure 3.6. *Principle for developing the profile used for route planning*

To assess our approach, we proceeded as in the previous section by using a set of problems from the initial Gendreau benchmark. We compare GA model solutions based on a dynamic planning profile where the GA model is based only on the static historical profile to evaluate the performance differences on the chosen objective function (i.e. sum of travel time + time delays with all customers). For this, the simulation is repeated for every solution provided on a set of actual speed profiles generated for each type of road. This generation is mainly based on historical velocity profiles. More precisely, the actual velocity will be composed of disruption phenomena and congestion whose definition is described below. A disturbance is a relatively small change in velocity; congestion is a significant change in velocity in the velocity history.

For example, in Figure 3.7 there is congestion in the first period followed by disruptions over the following periods of the day. We generated several actual velocity profiles with 0 to 3 congestions on the planning horizon.

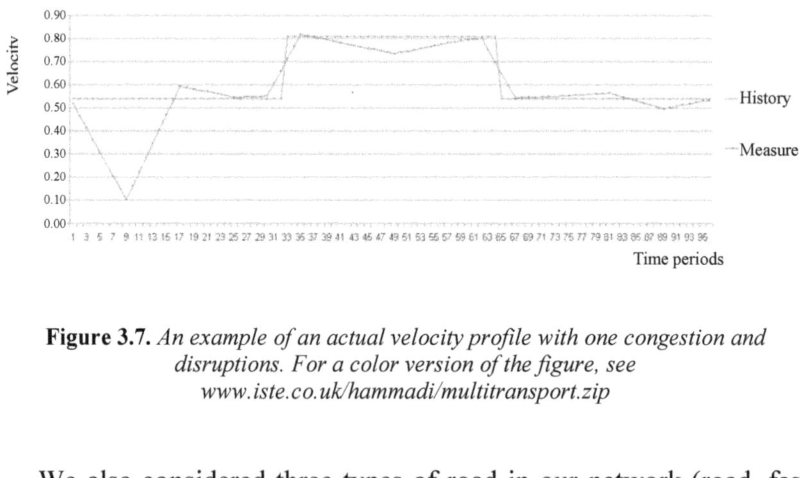

Figure 3.7. *An example of an actual velocity profile with one congestion and disruptions. For a color version of the figure, see www.iste.co.uk/hammadi/multitransport.zip*

We also considered three types of road in our network (road, fast road and highway). To do this, we created different actual velocity profiles for the three road types as mentioned in [ICH 03].

In our simulation platform, it is assumed that we receive traffic information (i.e. actual measured velocity) of a period corresponding to 1/24th of the opening time of the day of service. For an opening horizon of 8 hours, this amounts to receiving the actual measured velocity every 20 minutes. We set the time required to simulate the opening time of a day of service to 60 minutes, as the number of time periods to consider is large (96 periods in the actual velocity profile).

In Table 3.4, we give the overall results for all categories of test problems C1, R1, RC1, C2, R2 and RC2 by setting the number of congestions to one congestion on each road type and slowing down by a factor of 2/3 the velocity between peak and off-peak periods.

In this version of our platform, the objective function used is a compromise (denoted VO in Table 3.4) between the total travel time of each vehicle of the fleet (denoted TT) and the sum of customer delays (denoted TR).

The results obtained for the VO compromise show that the R1 resolution (obtained by the model based on a dynamic profile) is always better than the R2 resolution (obtained by the model based on historical patterns) for all categories of test problems. The level of improvement (i.e. percentage gain) varies between 5.48% and 23.08%, which confirms the value of a dynamic velocity profile based on real-time information and a predictive model.

	C1		R1		RC1	
	R1	R2	R1	R2	R1	R2
TT	1301.08	1202.31	1364.63	1367.85	1413.37	1433.21
TR	335.48	925.26	744.31	912.51	644.37	743.73
VO	1636.56	2127.57	2108.94	2280.36	2057.74	2176.94
Gain (%)		23.08		7.52		5.48
	C2		R2		RC2	
	R1	R2	R1	R2	R1	R2
TT	866.53	866.44	1403.42	1510.08	1477.25	1470.73
TR	229.33	497.45	1758.60	2232.45	594.55	896.95
VO	1095.86	1363.90	3162.02	3742.53	2071.80	2367.68
Gain (%)		19.65		15.51		12.50

Table 3.4. *Results of the 6 categories of problems for scenarios D2 and V1*

To conclude the work of this simulation platform, we have shown and were able to assess the interest in considering uncertainties and real-time information in dynamic management approaches to vehicle routing. The GA-type optimization algorithm has demonstrated its effectiveness in the dynamic model, but the architecture of the simulation platform can integrate other metaheuristic algorithms (tabu search, ant colony, etc.). The stochastic model with fuzzy requests and dynamic velocity profile model are independent of the optimization method chosen and can easily be integrated into other optimization approaches.

3.4. Real-time fleet management: a case study

Spatial technologies create, or make possible, real-time knowledge of the position of vehicles, as well as the exchange of information in different formats between the Dispatch Center (DC) and vehicle occupants, etc. They also provide the manager installed at the DC with the opportunity to access a large amount of information. To enable the manager to fully utilize this information, two methods of decision support are needed to help in the task of planning fleet activities. In the proposed application, all tasks performed using vehicles over a given time horizon are generally not defined. Tasks can either be unknown or require change or clarification. This uncertainty prevents the manager from performing a tactical plan of tasks. Generally, he must perform real-time planning.

More precisely, the DC manager faces two main problems. The first problem is that of assigning a new task or a modified task to a vehicle. The second problem is a consequence of the former. Following a change in the activity of a vehicle, even partially known tasks must be planned once again between the vehicles of the fleet in order to best complete them. This second problem goes beyond the new allocation of some of the tasks previously assigned to the vehicle to which the new task has been assigned. It involves a truly global vision and consideration of the exchange of tasks between the different vehicles in the fleet. These exchanges, however, are constrained by various practical constraints. These two problems have so far been little discussed because technology has not allowed access to the data needed to solve them.

In this case study, the objectives are, one the one hand, to develop decision support methods able to provide assistance for these two problems. These algorithms will be used in a context characterized by both the mass of data available and the real-time dimension. Support methods for real-time planning should have the following characteristics. They must be: i) flexible to adapt to different types of constraints; ii) robust in terms of quality of the solutions provided; iii) fast in providing a response with very limited computation time.

3.4.1. *General architecture of the optimization engine*

The first step was to design the optimization engine architecture to manage the fleet in real time.

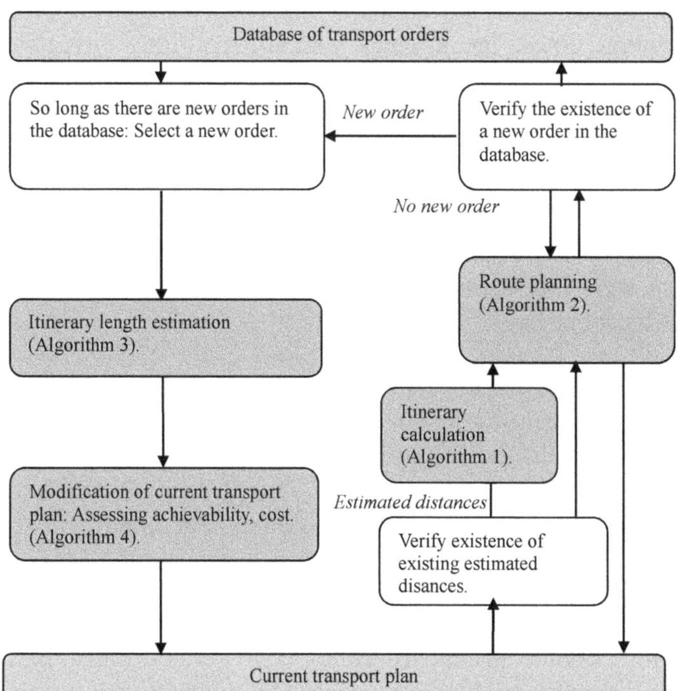

Figure 3.8. *Optimization engine*

This optimization engine, described in Figure 3.8, consists of two types of algorithms: algorithms related to the determination of itineraries (Algorithms 1 and 3), and algorithms related to route planning (Algorithms 2 and 4). Algorithms 1 and 3 are used when a new transport order arrives in order to immediately to assess the impact of this addition. Algorithms 2 and 4 are used when a new fleet activities plan is determined.

3.4.2. *Itinerary calculation and length estimation*

The problem of computing itineraries in a network is a classical operations research problem for which several algorithms have been proposed. The main difficulty we have to face is the amount of data available. The geographic database of the French road network leads to a directed graph of over 6 million nodes and 15 million arcs connecting them. Therefore a classic implementation of a shortest path algorithm such as the Dijkstra algorithm leads to prohibitive computation times, inconsistent with the real-time environment in which the optimization engine must evolve. For example, the computation of the shortest path between Bayonne and Toulouse (273 km) requires more than 1,000 seconds on a Pentium IV 1.2 Ghz. Let us note that this path was obtained without taking trucking restrictions into account. Therefore, we were led to develop a procedure for estimating the length of an itinerary (Algorithm 3) and modify the Dijkstra algorithm to make it more efficient in terms of computing time (Algorithm 1).

The procedure for estimating itinerary length is based on a linear regression between the length of the shortest path and the Euclidean distance between the origin and its destination. The regression parameter was estimated on a set of 91 path lengths. This procedure makes it possible to have a constant estimation of length.

For itinerary computation (Algorithm 1), we modified the Dijkstra algorithm by involving an estimate of the remaining length and achieving a double search: from the origin to the destination, as well as the destination to the origin. In addition, in the search we promoted the use of the main axes. Finally, we decomposed the network by distinguishing the main road network and a secondary network. These different modifications transform the exact algorithm into a heuristic approach. However, this resulted in good quality paths in very short computation times. Thus, for example, for Toulouse-Bayonne, we get a path of 296 km in 0.3 seconds.

3.4.3. *The static route planning problem*

After a study of the problem raised by the carrier, we were able to identify the class of problems to which it could be attached. In the static case, meaning when all the data describing the problem is known, it is a routing problem with pick-up and delivery constraints, time-window restricitions and a preset fleet size (PDVRPTW). More precisely, the PDVRPTW determines m tours such that:

(i) the route of truck k begins in k^+ and ends in k^-;

(ii) any transport order v is assigned to one and only one truck, the loading at v^+ is carried out before unloading at v^-;

(iii) the total load of truck k during the route must not exceed its capacity;

(iv) for each transport order v, truck loading in v^+ (respectively the truck unloading in v^-) takes place in the time slot $[e_v^+, l_v^+]$ (resp. in time slot $[e_v^-, l_v^-]$). For each truck k, the beginning of the route k^+ occurs in the slot $[e_s^{k+}, l_s^{k+}]$ and the end of the route k^- takes place in the slot $[e_s^{k-}, l_s^{k-}]$;

(v) the total cost of m routes is minimized.

A literature review of studies on problems of this type was initially carried out as well as a comparative study of different approaches to identify those likely to be used for real-time management.

This led us to propose, for the PDVRPTW, a generalization of a strong adaptive algorithmic approach developed for the vehicle routing problem with time windows. This method is an approximate solution algorithm based on the tabu search, which contains, amongst others, characteristics that involve a very small number of control parameters.

Different versions of the algorithm were developed in order to treat a broader problem than the PDVRPTW and improve performance in terms of quality of solutions produced relative to computation time. This involved, in fact, taking into account:

– social constraints such as breaks, daily working hours;

– a planning horizon over several days.

From an algorithmic point of view, the focus was on the development of:

– a preprocessing step based on the notion of compatibility between transport orders to reduce the solution space;

– a strategy to detect the possible infeasibility of an instance;

– adapted improvement procedures for modified routes.

On test instances involving 100 transport orders and a dozen trucks, computation times ranged between 2 and 65 seconds on a Pentium IV 1.2 Ghz, while solution values were at most 2.5% of the best known values.

3.4.4. *Route planning and modification of the transport plan*

Building on the work done on the static case, we developed Algorithms 2 and 4 in the engine optimization.

A modification in the transport plan can have several causes. It may be because of:

– the addition of a new set of transport orders;

– the withdrawal of a set of transport orders;

– delayed trucks;

– broken down trucks;

– new trucks available.

To assess the impact of a modification, we considered two approaches. The first was to run the route planning algorithm (Algorithm 4) for a limited computation time (see below for more details). This approach was examined by studying the quality of solutions produced according to the computation time. The second was to use a specific algorithm designed to build a good solution within a very small computation time. It is a constructive method which consists of identifying the first m transport orders to initialize routes, then to complete them by considering the orders according to their compatibility. In this insertion phase, results from the pre-treatment step mentioned above are used.

In case the modification addresses new transport orders, it should be noted that it is possible that no feasible solution exists.

Choosing the best approach depends on the strategy used by the carrier when faced with a change in the course of the route. The carrier may choose to either respond immediately or react immediately but negotiate, particularly on the time constraints (for example the addition of transport orders), or respond after a period of time.

The real-time route planning algorithm is based on an algorithm developed in the static case. It is distinguished by the fact that the algorithm must consider the previously approved route plans, including the fact that:

– certain transport orders have been completed;

– for some transport orders, collecting the goods has been completed;

– vehicles are at given positions.

Moreover, for vehicles in motion, a *decision period* corresponding to the time interval required for the validation of the new transport plan was defined. This period allows for route planning to be established, based not on current positions of vehicles but on projected ones. The static algorithm has hence been modified to deal with such a situation.

To validate these algorithms, we first developed a mini-simulator and a set of test instances. The purpose of this simulator was to reproduce the sequence of events that may occur during a period of time. Only the arrival of new transport requests was taken into account. The test instances were generated by considering the percentage of known orders ranging from 25% to 90%. This factor is a measure of the dynamic nature of the problem addressed. According to this factor, computational experiments showed that we accepted, because we were able to find a feasible solution, 65 to 90% of the total transportation orders; the number of new transport plans varying from 35 to 7. Finally, if 11 trucks were diverted from their original destinations, when 25% of orders were known, this was never the case when 90% of the orders were known.

3.5. Conclusion

The problem of real-time fleet management is a challenge that can be analyzed from several points of view:

– for carriers: to reduce costs, improve service quality, etc.;

– for the operational research community: to develop new methods to incorporate the real-time aspect;

– for the development of new computer and telecommunications equipment, etc.

Throughout this chapter, we have presented the levers on which actions are led and developed, both in the community of carriers and that of operational research, both of which have several areas of interest in this domain.

3.6. Bibliography

[BIA 00] BIANCHI L., Notes on dynamic vehicle routing - The state of the art, Technical Report IDSIA-05-01, 2000.

[BRA 01] BRANKE J., *Evolutionary Optimization in Dynamic Environments*, Kluwer Academic Publishers, December 2001.

[CHE 06] CHEN H.K., HSUEH C.F., CHANG M.S., "The real-time time-dependent vehicle routing problem", *Transportation Research Part E: Logistics and Transportation Review*, 42:383-408, 2006.

[COR 03] CORDEAU J.-F., LAPORTE G., "The Dial-a-Ride Problem (DARP): Variants, Modeling Issues and Algorithms", *4OR – Quarterly Journal of the Belgian, French and Italian Operations Research Societies*, 1:89-101, 2003.

[DRO 86] DROR M.., TRUDEAU P., "Stochastic vehicle routing with modified savings algorithm", *European Journal of Operational Research*, 23(2):228–235, 1986.

[DUB 88] DUBOIS D., PRADE H., *Possibility Theory: An Approach to Computerized Processing of Uncertainty*, Plenum, New York, 1988.

[GEN 96] GENDREAU M., LAPORTE G., SÉGUIN R., "Stochastic vehicle routing", *European Journal of Operational Research*, 88(1):3–12, 1996.

[GEN 99] GENDREAU M., GUERTIN F., POTVIN J.-Y., TAILLARD E., "Parallel tabu search for real-time vehicle routing and dispatching", *Transportation Science*, 33:381-390, 1999.

[GEN 01] GENDREAU M., LAPORTE G., SEMET F., "A dynamic model and parallel tabu search heuristic for real-time ambulance relocation", *Parallel Computing*, 27:1641-1653, 2001.

[GEN 06] GENDREAU M., GUERTIN F., POTVIN J.-Y., SÉGUIN R., "Neighborhood search heuristics for a dynamic vehicle dispatching problem with pick-ups and deliveries", *Transportation Research Part C: Emerging Technologies*, 14, 157-174, 2006.

[GOL 89] GOLBERG D.E., *Genetic Algorithms in Search, Optimization and Machine Learning*, Addison-Wesley, 1989.

[GON 07] GONCALVES G., HSU T., DUPAS R., HOUSROUM H., "Une plate-forme de simulation pour la gestion dynamique de tournées de véhicules", *Journal Européen des Systèmes Automatisés*, 41(5): 515-539, 2007.

[HOU 04] HOUSROUM H., HSU T., DUPAS R., GONCALVES G., "An evolutionary approach for the solving of the dynamic vehicle routing problem with time windows", *Conférence Internationale en Recherche Opérationnelle FRANCORO IV*, Fribourg, Switzerland, 18-21 August, 2004.

[ICH 03] ICHOUA S., GENDREAU M., POTVIN J.Y., "Vehicle dispatching with time-dependent travel times", *European Journal of Operational Research*, 144:379-396, 2003.

[LAR 00] LARSEN A., The dynamic vehicle routing problem, PhD Thesis, Lyngby, IMM-PHD-2000-73, Denmark, 2000.

[LEN 81] LENSTRA J.K., RINNOOY KAN A.H.G., "Complexity of vehicle routing and scheduling problem", *Networks*, 11: 221-227, 1981.

[LUN 96] LUND K., MADSEN O., RYGAARD J., Vehicle routing problems with varying degrees of dynamism, Technical Report, IMM, The Department of Mathematical Modelling, Technical University of Denmark, 1996.

[MAL 04] MALCA F., SEMET F., "A tabu search algorithm for a dynamic pickup and delivery vehicle routing problem", *Tristan V Triennal Symposium on Transportation Analysis*, 2004.

[MIT 04] MITROVIC-MINIC S., KRISHNAMURTI R., LAPORTE G., "Double-horizon based heuristics for the dynamic pickup and delivery problem with time windows", *Transportation Research Part B*, 38:669-685, 2004.

[PIL 12] PILLAC V., GENDREAU M., GUÉRET C., MEDAGLIA A.L., "A review of dynamic vehicle routing problems", *European Journal of Operational Research,* 225: 1-11, 2012.

[POW 95] POWELL W.B., JAILLET P., ODONI A., "Stochastic and dynamic networks and routing", in BALL M.O., MAGNANTI T.L., MONMA C.L., NEMHAUSER G.L. (eds), *Handbook in Operations Research and Management Science*, vol. 8, Network Routing, Elsevier, Amsterdam, pp. 141-296, 1995.

[PRI 04] PRINS C., "A simple and effective evolutionary algorithm for the vehicle routing problem", *Computers and Operations Research*, 31(12):1985–2002, 2004.

[PSA 95] PSARAFTIS H.N., "Dynamic vehicle routing: Status and prospects", *Annals of Operations Research*, vol. 61(1), pp 143-164, 1995.

[SOL 87] SOLOMON M., "Algorithms for the vehicle routing and scheduling problems with time window constraints", *Operations Research*, 35:254–265, 1987.

[TAI 97] TAILLARD E., BADEAU P., GENDREAU M., GUERTIN F., POTVIN J.-Y., "A tabu search heuristic for the vehicle routing problem with soft time windows", *Transportation Science*, 31:170–186, 1997.

[TAN 07] TAN K.C., CHEONG C.Y., GOH C.K., "Solving multiobjective vehicle routing problem with stochastic demand via evolutionary computation", *European Journal of Operational Research*, 177(2):813-839, 2007.

[TEO 96] TEODOROVIC D., PAVKOVIC G., "The fuzzy set theory approach to the vehicle routing problem when demand at nodes is uncertain", *Fuzzy Sets and Systems*, 82(3):307-317, 1996.

[THA 95] THANGIAH S.R., "Vehicle routing with time windows using genetic algorithms", *Handbook of Genetic Algorithms: New Frontiers*, 2: 253–277, 1995.

[WER03] WERNERS B., DRAWE M., "Capacitated vehicle routing problem with fuzzy demand" in VERDEGAY J-L. (ed.), *Fuzzy Sets Based Heuristics for Optimization*, Berlin, Springer, 2003.

[XU 08] XU J., GONCALVES G., HSU T., "Genetic algorithm for the vehicle routing problem with time windows and fuzzy demand", *IEEE World Congress on Computational Intelligence, WCCI 2008*, Hong Kong, China, June 1-6, 2008.

[YAN 04] YANG J., JAILLET P., MAHMASSANI H., "Real-time multi-vehicle truckload pick-up and delivery problems", *Transportation Science*, 38:135-148, 2004.

[ZAD 78] ZADEHL.A., "Fuzzy sets as a basis for a theory of possibility", *Fuzzy Sets and Systems*, 1:3–28, 1978.

[ZHA 09] ZHAO X., GONCALVES G., DUPAS R., "Un problème de tournées de véhicules temps réel avec temps de parcours variables", *Société Française de Recherche Opérationnelle et d'Aide à la Décision (ROADEF)*, 10th Conference, Nancy, France, 2009.

Chapter 4

Solving the Problem of Dynamic Routes by Particle Swarm

4.1. Introduction

One of the major concerns of companies whose operation depends on the transportation of products is to improve their distribution systems. Following the observable reduction of production costs in recent years, the share related to distribution costs has significantly increased relative to the cost of the product. Therefore, for many companies, lower distribution cost is a new challenge that can only be met through solving many combinatorial problems such as the development of vehicle routing. Nowadays, most transport systems must be able to operate within strict time constraints and adapt to the vagaries of life today.

Customers or partners in a business require a guaranteed quality of service (i.e. respected deadlines). In addition, the environment in which they operate is often uncertain and responsiveness is an important asset. This has led to the development of management models of transport systems, also called dynamic management, in

---

Chapter written by Mostefa Redouane KHOUAHJIA, Laetitia JOURDAN and El Ghazali TALBI.

which some of the data are considered to be time dependent. This is especially favorable because of recent advances in the field of ICT, so a fleet can be managed in real time.

When used together, devices such as geographic information systems (GIS), global positioning systems (GPS), sensors for traffic flow, and cell phones are capable of providing relevant, real-time data, such as the current location of a vehicle, traffic congestion, and periodic estimates of travel time. If this large amount of data is properly dealt with, it can be used to reduce costs and significantly improve the Quality of Service. Therefore, routes can be modified so as to better manage new events that disrupt a set route.

In this chapter, we study the dynamic extension of the conventional vehicle routing problem and its resolution by a recent metaheuristic paradigm. Introduced in 1995 by Kennedy and Eberhart [KEN 95], this paradigm is that of particle swarm optimization.

In section 4.2, we present the dynamic vehicle routing problem, its modeling and its challenges. In section 4.3, we introduce the metaheuristic particle swarms and, in section 4.4, describe its adaptation to resolve the problem being addressed. Section 4.5 presents numerical results and analysis. Section 4.6 closes with a conclusion and prospects.

4.2. Vehicle routing problems

In this section, we introduce in the first instance the vehicle routing problem in its static version, and then present the dynamic variation.

4.2.1. *The static vehicle routing problem*

The vehicle routing problem (VRP) is an optimization problem which was formulated by Dantzig and Ramser in 1959 [DAN 59]. It consists of determining, by minimizing the route cost, a set of routes for a limited number of vehicles, beginning and ending at a depot, such that each customer is visited exactly once by a vehicle, and the sum of the requests from customers on a route does not exceed the capacity of the vehicle that covers this route.

In a more formal manner, the VRP can be modeled as follows:

Let $G = (C, E)$ be a graph, where $C = \{ c_0, c_1, c_2.., c_n \}$ is a set of vertices with vertex c_0 taken as the depot, $E = \{(c_i, c_j)| c_i, c_j \in C \}$ is the set of arcs, and $C' = C \setminus \{c_0\}$ is the set of cities or customers.

Each arc is associated with a non-negative cost d_{ij}. This can be interpreted as the cost of travel or travel time between c_i and c_j.

We assume that we have m identical vehicles of the same capacity. The VRP consists of determining a set of routes of minimum cost (i.e. distance) such that:

– each city of C' is visited once and only once by exactly one vehicle;

– all routes begin and end at the depot;

– certain constraints are met. These constraints determine the different variants of the problem.

The problem can be expressed as follows:

$$\text{Minimize } \sum_{i=1}^{n} \sum_{j=1}^{n} d_{ij} \sum_{k=1}^{m} x_{ij}^{k}$$

where:

– n is the number of vertices;

– m is the number of vehicles;

– d_{ij} is the cost or distance between vertices c_i and c_j.

$$x_{ij}^{k} = \begin{cases} 1 & \text{if the segment } (c_i, c_j) \text{ is covered by vehicle } k, \\ 0 & \text{otherwise} \end{cases}$$

Among the various constraints, there are time constraints on the length of a route, or time windows, indicating the time intervals in which the services are to begin.

In the case of vehicle route planning in static mode, it is generally accepted that [LAR 00]:

– All information (i.e. data of the problem) for route planning is assumed to be known by the planner before the planning process begins.

– This information does not change after itineraries are made.

– This information includes all customer attributes such as geographical location, service time spent on-site, and each customer request (i.e. amount to be collected or delivered). In addition, information about the travel time required between the customers to be served must be known, or measurable, by the planner.

4.2.2. *The dynamic vehicle routing problem (DVRP)*

In general, the VRP is solved statically, by considering that all the information is known in advance and is reliable over time. However, in practice, it is possible that one or more contingencies disrupt previously established routing. In this case, the problem must be solved in real time. This appears as a dynamic vehicle routing problem (DVRP) [DAN 59].

In [LAR 00], Larsen defined the DVRP problem as a static VRP problem variant, differing due to the following features:

– Information about the route planning is not fully known by the planner when the planning process begins. In other words, some of the problem data are explicitly time-dependent. New commands can arrive during the course of the day of service, and must be integrated into the existing schedule.

– Some events can disrupt the existing plan and must be reconsidered to respond to the new context better.

It is obvious that the dynamic vehicle routing problem is more complex than the static problem. If the class of VRP problems is denoted by P(VRP) and the class of DVRP problems is denoted by P(DVRP), then P(VRP) $\subset$ P(DVRP). Since the problem of static VRP is known as a NP-complete problem, namely, that it is not possible to find an optimal solution in a reasonable computation time, then the dynamic vehicle routing problem DVRP also belongs to the class of NP-complete problems. This can be summarized in the resolution of each time slice of a partial static VRP. In Figure 4.1, we have a simple example of the dynamic VRP, where two vehicles must serve a set of customers.

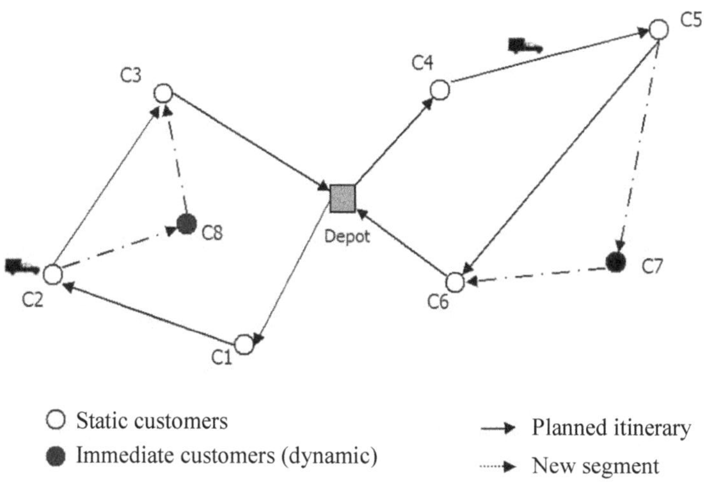

Figure 4.1. *Dynamic vehicle routing problem*

Static customers are represented by white nodes, while dynamic customers (i.e. new customers) are represented by black nodes. The arcs represent the planned itinerary for each vehicle. New customers must be inserted into the already-planned itineraries. The new itinerary segments are shown as a dashed line.

Unlike a static vehicle routing problem, dynamic management not only depends on the number and spatial distribution of past customers, but also on the number of immediate (dynamic) events and instances, when these events actually take place. A measure to describe the dynamics of the system is the degree of dynamism (*dod*). It is described as follows [LAR 00]:

$$dod = \frac{Number\ of\ dynamic\ customers}{Total\ number\ of\ customers}$$

A static routing problem can be distinguished by a *dod* of nil, while a purely dynamic problem has a *dod* equal to 1.

4.2.3. *Importance of dynamic routing problems*

There are several problems that must be solved in real time. In what follows, we present the main applications that motivate the research into the VRP in realtime [GHI 03]:

– *Dynamic fleet management:* several trucking operations require large-scale, real-time transmission of vehicles in order to collect or deliver cargo. Significant savings can be achieved by optimizing these operations.

– *Large distribution networks:* distribution companies assess the level of the inventory in order to renew their stock before it is sold out. As a result, requests are known in advance and all customers are static. However, because the request is uncertain, some customers (usually a small percentage) may be short of reserves, and must be constantly served.

– *Letters:* long-distance postage requires collecting parcels locally before sending them to a remote terminal. This collection allows the consolidation of loads. In addition, cargo from remote terminals must be distributed locally. Most pickup requests are dynamic and must be served, if possible, on the same day.

– *Towing and repair companies:* companies that provide services for towing and repair (broken-down car repair, equipment repair, etc.) are constantly inundated with urgent requests that must be met during the service.

– *On-demand transport systems (dial-a-ride systems):* on-demand transport systems provide transportation services to people by linking them between given destinations. Customers can book their trip a day in advance (static customers) or make a request at the last minute (dynamic customers).

– *Emergency services:* emergency services include the police, the fire service and the ambulance service. By definition, all customers are dynamic. In addition, the request rate is generally low, which leads to vehicles being unoccupied most of the time. In this context, repositioning vehicles to avoid congestion in the city center during rush hour is a major issue.

– *Taxi services:* in taxi services, almost all customers are dynamic. As in the case of emergency services, temporary relocation of unoccupied vehicles is a response to this dynamism.

4.3. Resolution scheme of the dynamic vehicle routing problem

There are two main components in the proposed resolution method:

1. The event planner collects new orders received during the day of service. It uses this information to build problems (instances) as partial static VRP. In addition, it manages the organization of assignments sent to vehicle drivers.

2. The APSO component (Adaptive Particle Swarm Optimization) is used as a heuristic for solving VRP partial instances. These instances are generated in advance by the event planner.

This resolution scheme was used by Motemanni in [MON 05], and also by Hoursoum [HOU 06] and Hanshar [HAN 07].

In what follows, we describe these two components.

4.3.1. *Event planner*

The event planner acts as an interface between the arrival of new requests and the optimization algorithm. Based on the principle of partitioning the day of service into several time periods (slices), as well as T_{co}, the "cut-off time" of new orders, the planner generates partial instances and sequentially runs the APSO algorithm on these instances. The solutions found during the search will be used to schedule the assignment of vehicles to serve customers.

The first static problem generated (at the beginning of the planning horizon) consists of all the requests that could not be served the previous day. The T_{co} parameter indicates the time until which requests can be processed. Beyond that time, requests are postponed to the following day of service. In other words, the optimization process begins with requests that have not been addressed during the previous day's service. Part of the subsequent problem includes all requests that have happened during the last time slice.

Customers begin to be served upon the arrival of vehicles. The loading capacity of each vehicle is equal to the residual capacity of all deliveries made throughout a day. At the end of each time slice, the solution found for the partial static problem is examined, and the requests that can be served during the next time slice are assigned to different vehicles.

Moreover, the vehicles return to the depot if and only if:

– all customers are served before the end of the day of service;

– the vehicle has used all its capacity.

In practice, a vehicle waits at the site of its last customer if neither of the two previous conditions is met. The routes may change with the arrival of new requests, and vehicles will therefore be reassigned to serve these requests.

The event planner pseudo-algorithm is presented in Algorithm 4.1. Initially, all *UnservedOrds* contain requests known from the previous day. The *Time* variable is initialized to 0, and vehicles are positioned at the depot. The simulation process is carried out in conjunction with that of the resolution. A partial instance (*PartialProb*) containing the *UnservedOrds* requests is generated and solved using the algorithm described in section 4.4.3. Therefore, vehicles are assigned to serve customers in accordance with the *PartialProb* instance solution, and requests can be processed during the time interval $[Time; Time + T_s[$.

Finally, it updates all *UnservedOrds* of unserved requests, positions, and residual capacity of vehicles. These operations are repeated until all customers have been served ($UnservedOrds = \varnothing$) and vehicles are assigned to the depot as their final destination.

Algorithm 4.1. Event planner procedure

- $Time := 0$;
- Vehicles are positioned at the depot;
- $UnservedOrds :=$ known requests from the previous day (all orders emitted after T_{co});

So long as ($UnservedOrds \neq \varnothing$) **do**

- $PartialProb :=$ instance with orders in $UnservedOrds$;
- Execute APSO on $PartialProb$;
- $CommOrds :=$ requests with time window $P_t \in [Time, Time + T_s[$;
- Assign vehicles service requests of the set $CommOrds$;
- $UnservedOrds := UnservedOrds \setminus CommOrds$;
- $UnservedOrds := UnservedOrds \cup \{$requests arrived during the last time slice$\}$;
- $Time := Time + T_S$;
- Update positions, loading capacity and journey times of vehicles;

End so long as

- Command vehicles to return to depot;

4.3.2. *Particle swarm optimization*

In this section, we present the basic concept of the particle swarm optimization (PSO) metaheuristic.

The particle swarm optimization [KEN 95] comes from an analogy of the collective behaviors of animal movements. In some groups of animals, such as schools of fish or flocks of birds, we can observe relatively complex movement dynamics, while individuals themselves only have access to limited information, such as the position and speed of their nearest neighbors.

These collective behaviors fit in perfectly with the theory of self-organization. To summarize, each individual uses the local information that they can access on the movement of its closest neighbors to decide its own displacement. Very simple rules, such as "stay relatively close to others", "go in the same direction", "go at the same speed" suffice to maintain the cohesion of the whole group, and allow complex and adapted collective behavior.

The authors of the particle swarm optimization method were inspired by this model of putting socio-psychology theory into perspective on information processing and decision making in social groups. The method involves large groups of particles in the form of vectors moving in the search space. Each particle i is characterized by its position $\vec{x}_i$, and a position change vector (called velocity) $\vec{v}_i$. At each iteration, the particle moves. The velocity is updated based on the socio-cognitive map of the swarm. The particles' displacement is governed by their past behavior and that of their neighbors (neighbors in the social network and not necessarily in space). During updates, the position of particles, direction of their displacement (their velocity), best previous position $\vec{p}_i$ and best position $\vec{p}_g$ amongst neighbors are taken into account. The change of position is as follows:

$$\vec{v}_i(t+1) = \vec{v}_i(t) + \varphi_1 \times r_1(\vec{p}_i - \vec{x}_i(t)) + \varphi_2 \times r_2(\vec{p}_g - \vec{x}_i(t)) \quad [4.1]$$

$$\vec{x}_i(t+1) = \vec{x}_i(t) + \vec{v}_i(t+1) \quad [4.2]$$

where: the parameters φ_1, φ_2 are variables drawn from $U_{[0,\varphi_{max}]}$ which have the role of weighing the relative roles of individual experience (φ_1) and social communication (φ_2). The random variables r_1 and r_2 take values from the interval [0, 1].

To prevent particles from moving too quickly from one region to another in the search space, the velocity is limited by V_{max} in each dimension.

The pseudo-code for the most general version is presented in Algorithm 4.2.

Improvements have been made to the basic algorithm, especially in terms of controlling convergence. In the basic version, the V_{max} parameter prevents the system from oscillating too much in the search space. Using *constriction coefficients* helps to better control this behavior [CLE 02]. In the same vein, another version implements an *inertia weight* by multiplying the velocity in the velocity equation by this coefficient. To summarize, the inertia weight decreases over time, causing a convergence controlled by this parameter. The range of movement of the particles decreases until it disappears.

Another important parameter is the notion of neighborhood or particle swarm topology. It seems to be commonly accepted that a circular social neighborhood produces better results than a star or radius social neighborhood.

Algorithm 4.2. Particle swarm optimization

For each particle i do

–Randomly initialize the current position $\vec{x}_i$;

–Initialize the best position of the particle $\vec{p}_i$ by its current position $\vec{x}_i$

$\vec{p}_i := \vec{x}_i$;

–Randomly initialize each dimension of $\vec{v}_i$ within a maximum limit [-V_{max}, V_{max}]

–Update the best position in the vicinity of the particle $\vec{p}_g := argmin\{f(\vec{p}_i)\}$

Repeat

For each particle *i* **do**

–Update position $\vec{x}_i$ according to equations (4.1) and (4.2);

If $f(\vec{x}_i) < f(\vec{p}_i)$ **then**

$\vec{p}_i := \vec{x}_i$;

End if

If $f(\vec{p}_i) < f(\vec{p}_g)$ **then**

$\vec{p}_g := \vec{p}_i$;

End if

End for

Until stop criterion is reached

4.4. Adaptation of the PSO metaheuristic for the dynamic vehicle routing problem

4.4.1. *Representation of particles*

The choice of representation is crucial to the success of an algorithm. In our approach, we use a representation of variable length, which includes information on customers who have already been visited and those whose service has already been planned.

The goal is to find a translation between the solution and the particle. In our work, we propose a representation dedicated to this kind of problem. The advantage of this representation in the paradigm used is that it maintains the discrete aspect of the problem, although generally most of the work is part of a translation problem subject to continuous variables. Importantly, the representation must allow the insertion of immediate (dynamic) customers on already-planned routes without affecting the partially completed routes.

The resolution of the problem includes a simulation component in which information about customers and vehicles is stored in real time. As the problem is dynamic, i.e. customer requests arrive over time, it is necessary to have knowledge of the status of each customer (served/not served) and the time of service. The simulation framework also stores information about the vehicle such as their position in the covered area, the remaining load capacity and the distance already traveled. Vehicles do not leave the depot at the same time, and are assigned to customers at the beginning of each time slice. Service times provide a scheduling simulation engine. One solution corresponds to a set of itineraries, each summarized in a permutation of customers to serve, as follows:

$$\mathrm{R_k} : (c_0, c_1, c_2, \ldots, c_i, \ldots, c_n, c_{n+1})$$

A route begins and ends at the depot ($c_0 = c_{n+1}$). For the simulation, we assign each customer c_i, the following information:

(X_i, Y_i): coordinates of customer c_i.

S_i: Boolean variable indicating whether customer c_i is served or not.

T_i: service time of customer c_i (time at which they were served).

The route R_k is served by vehicle v_j. For each vehicle v_j, we keep the following information:

– (X_j, Y_j): the coordinates of vehicle v_j in the covered area;

– Cap_j: the remaining capacity of vehicle v_j;

– $Dist_j$: the distance covered by vehicle v_j.

The initial population is obtained by generating a random permutation of static customers defined by the T_{co} parameter (cut-off time of new requests). The solution changes at every arrival of a new request, and increases by one dimension for each new customer.

4.4.2. *Velocity and movement of particles*

As our representation is based on a permutation of customers, a simple strategy for updating the velocity vector and the movement of particles is needed. For this purpose, we consider inertia as a constant and equal to 1.

The coefficients φ_1, φ_2 control the attraction of the best individual performance and best overall performance, respectively. According to probabilities r_1 and r_2, particles are guided to the best position $\vec{p}_i$, or the best positionin the swarm $\vec{p}_g$. The movement of particles is summarized by the application of the λ-exchange operator with motion (1, 0). This operator is used to select a non-served customer and re-insert them into another route with a strategy of least cost. This is described in Figure 4.2. Any operator is applied so that the order in which a served customer appears cannot be changed. In addition, its application requires the calculation of the cost of insertion, within the constraints related to the feasibility of the solution (vehicle load, closure of depot, etc.). In every generation, the particles move by applying the operator.

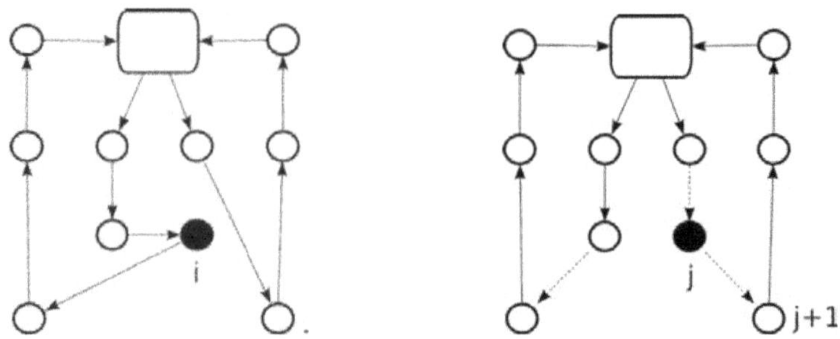

(a) (b)

Figure 4.2. *Insertion operator: (a) selection, (b) replacement*

The optimization process is repeated at each time slice, and a new partial instance is solved. The size of the position vector increases dynamically with the arrival of each new request.

Our metaheuristic was hybridized with a local 2-Opt search [LIN 65], which consisted of uncrossing a route by selecting two arcs $a[i_1, i_2]$, and $b[j_1, j_2]$, and replacing them with $a'[i_1, j_1]$ and $b'[i_2, j_2]$. Figure 4.3 illustrates the application of 2-Opt on a route. In our approach, the local search is applied after each movement of particles.

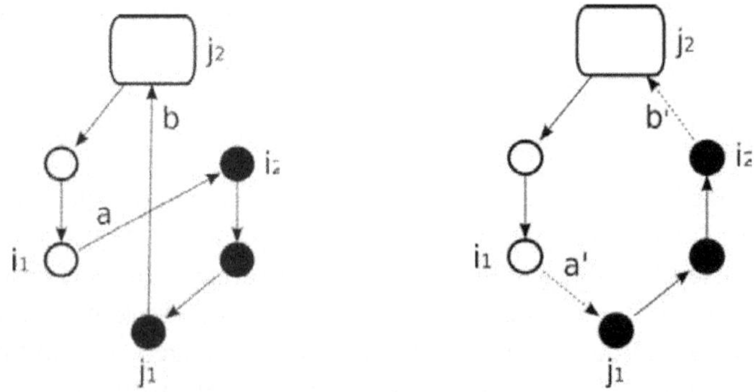

Figure 4.3. *Application of the 2-Opt operator*

4.4.3. *The APSO algorithm (Adaptive Particle Swarm Optimization)*

The algorithm is slightly modified to that of the canonical PSO, in order to take the dynamic aspect of the problem into account.

At each time slice, the position of each particle is initialized from adaptive memory. New requests are inserted into this new position. The process is repeated until a stop condition is met. In our case, it is set to a predefined number of iterations. The APSO pseudo-code for the DVRP problem is presented in Algorithm 4.3.

Algorithm 4.3. Adaptative optimization by particle swarm for the DVRP

INPUT: DVRP partial instance I_j($j\in[0,..,n]$) corresponding to the set of all customers who arrived during the previous time sliceT_j;
For each partial instance I_j **do**
  For each particle *i* **do**
  // Insert new requests
  $\vec{x}_i :=$ Insert ($\vec{x}_i$, *New requests*);
  $\vec{p}_i :=$ Insert ($\vec{p}_i$, *New requests*);
  Evaluate ($\vec{x}_i$, $\vec{p}_i$);
  If $f(\vec{x}_i) < f(\vec{p}_i)$ **then**// Re-use best solutions found by the swarm
    $\vec{p}_i := \vec{x}_i$;
  Otherwise
    $\vec{x}_i := \vec{p}_i$;
  End if

  End for

  Repeat
    For each particle *i* **do**
    - Update position $\vec{x}_i$ according to Eqs. (4.1) and (4.2)
    - Evaluate $f(\vec{x}_i)$;
    If $f(\vec{x}_i) < f(\vec{p}_i)$ **then**
    $\vec{p}_i := \vec{x}_i$ // Update best position of particle
    End if

If $f(\vec{p}_i) < f(\vec{p}_g)$ **then**

$\vec{p}_g := \vec{p}_i$; // Update best position in the swarm;

End if

End for

Until continuation criterion

End for

4.4.4. *Adaptive memory mechanism*

When solving the problem using APSO, each particle memorizes the best position it has achieved during its search. The set consists of the best positions found by the swarm, which represents an additive memory to the system. This memory is adaptive because it changes at each time slice in order to better respond to changes in the environment. These positions are summarized by the solutions found, namely the customers visited in sequence and the partial routes that minimize the distance traveled.

The concept of adaptive memory has been successfully applied to the dynamic vehicle routing problem by integrating it into a tabu search proposed by Gendreau in [GEN 99].

Each particle has a set of solutions combining its current position and the best position it could find. The main idea is that when there is a change in environment, the particle should be repositioned to the position that responds best to the new context (partial problem) in terms of solution quality.

The main idea of adaptive memory is that the memorization of previous solutions may have some advantage in a changing environment. When faced with dynamic problems, it is rare that the solution remains unchanged between two time periods. Moreover, usually in the DVRP case, the new overall optimum appears near the vicinity of the previous solution. Thus, the previous optimum can be re-used to more effectively monitor the evolution of the overall optimum. If there is a new change in the environment and the overall optimum is in the vicinity of the previous optimum (both problems are potentially very similar), the search for the newer optimum will be more effective.

Updating the adaptive memory is performed by the following steps:

1. If the memory is not empty: build an initial solution from routes already in the memory, and initialize the current position as the starting solution.

2. Insert new customers into the starting solution. These customers are those who have issued a request for service during the last time slice.

3. Apply the PSO algorithm. The memory is updated at each iteration.

Finally, our simulation model can control the dynamics of the environment. The idea of partitioning the day of service into multiple time slices overcomes the problem known as "obsolete memory". Consequently, the algorithm is reactive and follows the overall optimum in its evolution perfectly.

4.5. Experimental results

4.5.1. *Datasets*

We used the benchmarks proposed by Kilby [KIL 98]. They are derived from conventional VRP benchmarks [TAI 93; CHR 84; FIS 81]. These instances include 50 to 199 customers characterized by their spatial coordinates, their requests, and the time needed to serve them. The number of customers can be deduced from the name of the instance. For example, the file "tai75a.vrp" is a Taillard modified instance [TAI 93], which has 75 clients. In [KIL 98], Kilby adapts these benchmarks by adding the time of occurrence of each customer order. However, the number of vehicles is set to 50. Capacity is uniform across the fleet. Velocity is set to unit of distance / unit of time. Instances discussed in this work are of the pick-up type. However, for all the benchmarks, a delivery variation is also available.

These benchmarks include three classes of instances: Taillard (*Tai-series*) [TAI 93], Christophides (*C-series*) [CHR 84] and Fisher (*F-series*) [FIS 81]. Each class has a different coverage area. Table 4.1 summarizes the characteristics of each class of benchmarks based on the density of customers to serve.

Type	Customers	Distribution
F-series	71-134	The majority of customers are grouped around a same region
Tai-series	50-199	Alternating between uniformity and grouping
C-series	75-150	Uniform

Table 4.1. *Details of topology and coverage areas*

The cut-off time of T_{co} orders is set to 0.5. This value leads to a degree of dynamism, *dod*, approximately equal to 0.5 for all processed instances.

The day of service is divided into 25 time slices of equal duration. This partitioning was defned after an experimental study led by Montemanni *et al.* in [MON 05] on the quality of solutions depending on the arrival of customer requests.

4.5.2. *Experiments and analysis*

Our approach was implemented over the ParadisEO platform (PARAllel and DIStributed Evolving Objects) [TAL 09]. This software platform is written in C++ and is dedicated to the re-design and implementation of single and multi-objective metaheuristics. It appears as an extension of the EO platform. It contains a wide range of metaheuristics based on unique solution metaheuristics (tabu search, simulated annealing, etc.) to population-based metaheuristics (genetic algorithms, particle swarm optimization, etc.). It also offers parallel algorithm models (island model, farmer-workers, etc.). Moreover, it provides a simple interface for the deployment and scaling of these models on computational grids. The implementation conducted in this work includes the event planner, as well as the APSO-DVRP component.

The comparison of the quality of solutions in terms of minimizing the covered distance is between APSO-(2-Opt) and different metaheuristics: Genetic Algorithm (GA)-(2-Opt), Tabu Search (TS) proposed by Hanshar *et al.* [HAN 07], and Ant System (AS) proposed by Montemanni *et al.*[MON 05].

The parameters chosen for our algorithm are:

– population size: 100;

– inertia weight $\omega = 1$;

– coefficients φ_1, φ_2 take values from the [0.5, 1.0] interval;

– the number of evaluations per time slice is equal to 100.

In Table 4.2, the solutions obtained by hybridized APSO are compared with those from the algorithms mentioned above.

The best solution (Min), and the average of the solutions (Average) are reported after 30 executions of the algorithm. Indicated in bold are the best solutions for each instance. The "Ratio" column shows the improvement brought about by APSO (positive), and the approximation error (negative) made by our algorithm for each instance.

Table 4.2 shows that our hybridized algorithm improves 4 of the 21 best solutions known from the literature. An improvement of 1.14%, 2.09%, 0.25% and 0.37% was recorded for instances c75, c120, f71, tai75b. In addition, 14 of the 21 solutions obtained by hybridized APSO are better and more competitive than those obtained by AS.

	Metaheuristics								
Benchmarks	APS-(2-Opt)		AS[1]		GA-(2-Opt)[2]		TS		Ratio(%)
	Min	Average	Min	Average	Min	Average	Min	Average	-
c50	575.889	591.208	631.3	681.86	**570.89**	593.42	603.57	627.9	-0.876
c75	**970.446**	*1067.53*	1009.36	1042.39	981.57	1013.45	981.51	1013.82	+1.146
c100	988.269	*1091.26*	973.26	1066.16	**961.1**	987.59	997.15	1047.6	-2.827
c100b	924.319	977.009	944.23	1023.6	**881.92**	900.94	891.42	932.14	-4.808
c120	**1276.88**	1450.82	1416.45	1525.15	1303.59	1390.58	1331.22	1468.12	+2.092
c150	1446.2	1618.21	1345.73	1455.5	1348.88	1386.93	**1318.22**	1401.06	-9.709
c199	1962.39	2036.62	1771.04	1844.82	**1654.51**	1758.51	1750.09	1783.43	-18.609
f71	**279.519**	323.18	311.18	358.69	301.79	309.94	280.23	306.33	7.968
f134	15875	16427.88	**15135.51**	16083.56	15528.81	15986.84	15717.9	16582.04	-4.886
tai75a	1816.07	1992.29	1843.08	1945.2	1782.91	1856.66	**1778.52**	1883.47	-2.111
tai75b	**1447.39**	1457.48	1535.43	1704.06	1464.56	1527.77	1461.37	1587.72	+1.186
tai75c	1481.35	1503.2	1574.98	1653.58	1440.54	1501.91	**1406.27**	1527.72	-5.339
tai75d	1414.28	1466.73	1472.35	1529	**1399.83**	1422.27	1430.83	1453.56	1.03
tai100a	2249.84	2394.47	2375.92	2428.38	**2232.71**	2295.61	2208.85	2310.37	-0.767
tai100b	2238.42	2388.38	2283.97	2347.9	**2147.7**	2215.93	2219.28	2330.52	-4.224
tai100c	1534.85	1606.22	1562.3	1655.91	1541.28	1622.66	**1515.1**	1604.18	-1.304
tai100d	1955.06	2059.7	2008.13	2060.72	**1834.6**	1912.43	1881.91	2026.76	-6.56
tai150a	3426.91	3535.16	3644.78	3840.18	**3328.85**	3501.83	3488.02	3598.69	-2.86
tai150b	3263.03	3469.61	3166.88	3327.47	2933.4	3115.39	**3109.23**	3215.32	-4.947
tai150c	2842.13	3137.73	2811.48	3016.14	**2612.68**	2743.55	2666.28	2913.67	-8.782
tai150d	3375.16	3571.73	3058.87	3203.75	**2950.61**	3045.16	2950.83	3111.43	-14.389

Table 4.2. *Results obtained by APSO-(2opt), AS-(2-Opt), GA, and TS*

4.5.3. *Measure of dynamicity*

In the context of the optimization of dynamic problems, a metaheuristic is an approach whose goal is not only to find the overall optimium after a specified number of generations, but rather an

1 [MON 05]
2 [HAN 07]

approach that can adapt and adjust to perpetual changes occurring in the environment.

The feature most often used to describe the dynamics of adaptivity of the algorithm is solution accuracy. The accuracy at time *t* for optimization algorithm A, and fitness function F, is defined as follows [WEI 03]:

$$accuracy^t_{F,A} = \frac{Min^t_F}{F(best^t_A)}$$

where: $best^t_A$ is the best candidate solution of the population at time *t*, and Min^t_F is the shortest distance traveled by the fleet vehicle that is known in the literature for the given problem. The accuracy varies between 0 and 1. When accuracy is equal to 1, the best solution from the population corresponds to the known optimum.

Benchmarks	Min^T_F	$F(best^{(T)}_A)$	Accuracy
c50	521	575.889	0.905
c75	830	970.446	0.855
c100	817	988.269	0.827
c100b	820	924.319	0.887
c120	1034	1276.88	0.810
c150	1028.42	1446.2	0.711
c199	1291.45	1962.39	0.658
f71	237	279.519	0.848
f134	1162.96	15875	0.073
tai75a	1618.36	1816.07	0.891
tai75b	1344.64	1447.39	0.929
tai75c	1291.01	1481.35	0.872
tai75d	1365.42	1424.91	0.958
tai100a	2041.336	2249.84	0.907
tai100b	1940.61	2238.42	0.867
tai100c	1406.2	1534.85	0.916
tai100d	1581.25	2001.62	0.790
tai150a	3055.23	3430.35	0.891
tai150b	2656.47	3263.03	0.814
tai150c	2341.84	2842.13	0.824
tai150d	2645.39	3375.16	0.784

Table 4.3. *Accuracy of APSO-(2-Opt) solutions on different dynamic instances*

From Table 4.3, the hybridized APSO algorithm gives solutions that have good approximations to those known from the literature in the static solutions' case[3]. The best solutions found in the dynamic case for instances *c75*, *c120*, *f71*, and *tai75b* have an accuracy ranging between 0.810 and 0.929.

In addition, we plotted the fitness variation for the best particle swarm for instance c75. Figure 4.4 shows this variation for each time period that includes the day of service. We can observe that the algorithm reacts perfectly to changes in the environment. As new requests arrive, we recorded a peak corresponding to a significant increase in the fitness of the particle, which is justified by the emergence of new segments or routes. Therefore, the APSO algorithm reacts immediately and starts its search to find the new optimum at each time slice. We can observe that our algorithm is very responsive to changes in the environment. This behavior gives it the ability to monitor the overall optimum. It demonstrates the effectiveness of the proposed approach by the self-organization aspect and adaptability that it offers to solve dynamic problems.

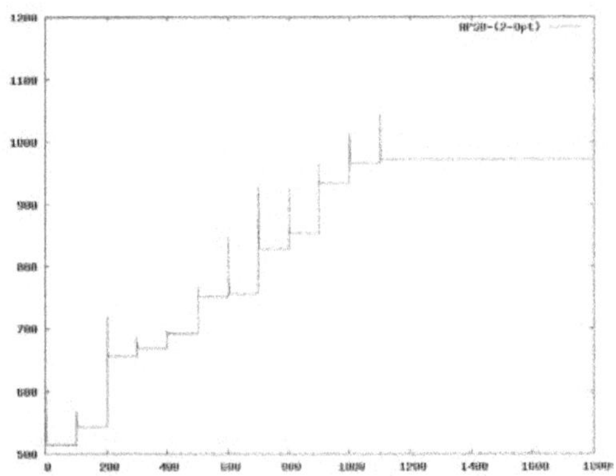

Figure 4.4. *Fitness variation for the best particle for dynamic instance c75*

Figure 4.5 illustrates roads built by our algorithm on instances representing the three types of coverage area as shown in Table 4.1.

3 http://branchandcut.org.

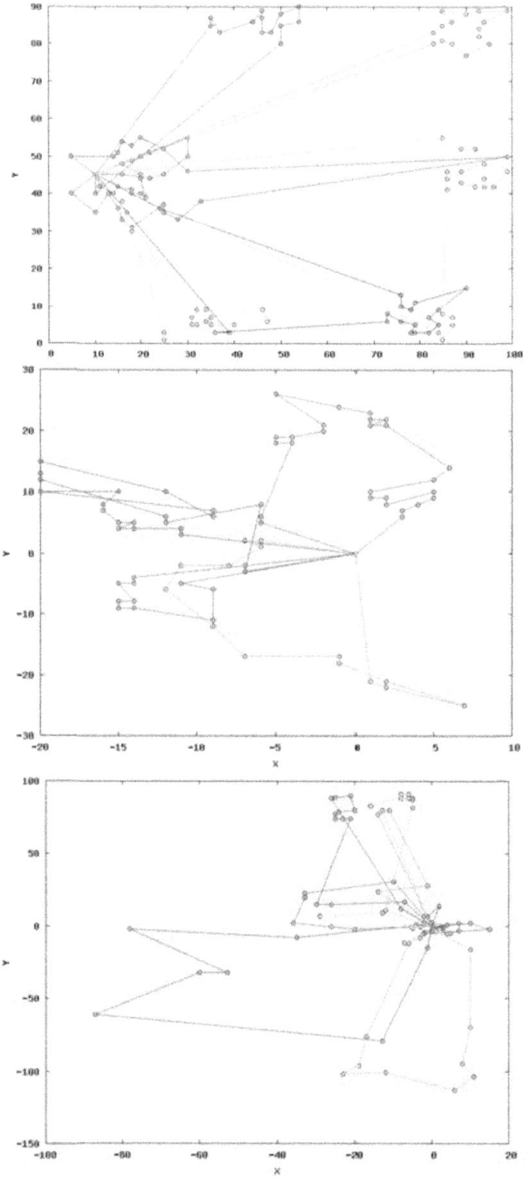

Figure 4.5. *Routes constructed by the APSO algorithm on instances f71, tai75b and c120, respectively. For a color version of the figure, see wwww.iste.co.uk/hammadi/multitransport.zip*

4.6. Conclusion

In this chapter, we presented the resolution of the dynamic vehicle routing problem by a metaheuristic of PSO. This paradigm is characterized by its adaptability and great flexibility. The proposed representation is generic and directly usable by other metaheuristics. To better respond to the dynamic nature of the problem, we have adapted this metaheuristic to the addressed problem. To do this, we used the concept of adaptive memory. This concept allows us to re-use the best solutions obtained by particles in order to build new solutions which will form the starting solutions for future changes in environment.

This approach was tested using problems from the literature. Our APSO approach produces very competitive solutions and outperforms conventional metaheuristics in different instances by offering better solutions. In addition, we measured the accuracy of solutions obtained dynamically with those obtained statically. Good approximation (accuracy) shows a reactive and adaptive behaviour from the algorithm. Regarding the prospects of our work, we will parallelize and scale our algorithm. A possible parallelization scheme is that of the multi-swarm model. This model makes several swarms cooperate by exchanging information and evolving in parallel in order to better respond to changes in the environment.

Finally, in this work we have developed a paradigm for solving dynamic vehicle routing problems. Through its genericity, it can be applied to different variants of the problem. This work certainly opens the door to other methods of resolution which will be based on the proposed paradigm.

4.7. Bibliography

[CHR 84] CHRISTOFIDES N., BEASLEY J. "The period routing problem", *Networks*, 14: 237–256, 1984.

[CLE 02] CLERC M., KENNEDY J., "The particle swarm-explosion, stability, and convergence in amultidimensional complex space", *IEEE Transactions on Evolutionary Computation,* 6 (1): 58–73, 2002.

[DAN 59] DANTZIG G.B., RAMSER J.H., "The truck dispatching problem", *Operations Research, Management Sciences*, 6 (1): 80–91, 1959.

[FIS 81] FISHER M.L., JAIKUMAR R., "A generalized assignment heuristic for vehicle routing", *Networks*, 11 (2), 1981.

[GEN 99] GENDREAU M., GUERTIN F., POTVIN J.Y., TAILLARD E., "Parallel tabu search for real-time vehicle routing and dispatching", *Transportation Science*, 33 (4): 381–390, 1999.

[GHI 03] GHIANI G., GUERRIERO F., LAPORTE G., MUSMANNO R.,"Real-time vehicle routing: solution concepts, algorithms and parallel computing strategies", *European Journal of Operational Research*, 15: 1–11, 2003.

[HAN 07] HANSHAR F.T., OMBUKI-BERMAN B.M., "Dynamic vehicle routing using genetic algorithms", *Applied Intelligence*, 27: 89–99, 2007.

[HOU 06] HOURSOUM H., HSU T., DUPAS R., GONCALVES G., "A hybrid GA approach for solving the dynamic vehicle routing problem with time windows", *2nd International Conference on Information & Communication Technologies: Workshop ICT in Intelligent Transportation Systems*, *ICTTA'06*, Damascus, Syria, April 2006

[KEN 95] KENNEDY J., EBERHART R., "Particle swarm optimization", *Proceedings of IEEE International Conference on Neural Networks*, vol. 4, Piscataway, NJ: IEEE 1942–1948, 1995.

[KIL 98] KILBY P., PROSSER P., SHAW P., Dynamic VRPs: a study of scenarios, Report APES-06-1998, University of Strathclyde, UK, 1998.

[LAR 00] LARSEN A., The Dynamic Vehicle Routing Problem, PhD thesis, Technical University of Denmark, 2000.

[LIN 65] LIN S., "Computer solutions of the traveling salesman problem", *Bell Systems Technical Journal*, 44: 2245-2269, 1965.

[MON 05] MONTEMANNI R. , GAMBARDELLA L.M., RIZZOLI A.E., DONATI A.V.,"A new algorithm for a dynamic vehicle routing problem based on ant colony system", *Journal of Combinatorial Optimization,* 10: 327–343, 2005.

[TAI 93] TAILLARD É., "Parallel iterative search methods for vehicle routing problems", *Networks*, 23 (8), 1993.

[TAL 09] TALBI E-G., *Metaheuristics: From Design to Implementation*, Wiley, New York, 2009.

[WEI 03] WEICKER K., “Performance measures for dynamic environments”, *Lecture Notes in Computer Science*, pp. 64–76, 2003.

Chapter 5

Optimization of Traffic at a Railway Junction: Scheduling Approaches Based on Timed Petri Nets

5.1. Introduction

Implementation of new high-speed European railways (TGV, particularly in Northern Europe) results in a significant increase in traffic density, which can be critical in certain places (railway junctions). This type of problem leads to the study of optimization scenarios for moving objects within a transport system. Typically, in a railway system, one wishes to determine itineraries of trains in order to optimize given criteria relating to traffic. It appears that this class of problem is closely related to scheduling problems. In particular, the optimization problem at a railway junction can be seen as a special case of the job-shop problem with blocking.

From another point of view, allocation of repetitive scenarios (daily bus, subway or train schedules) is a form of cyclic scheduling. These two applied themes are generating increasing interest from several teams in the Nord Pas de Calais region in France, (LAMIH/ROI, LAGIS/SED, LABOGP), and have been the subject of

---

Chapter written by Thomas BOURDEAUD'HUY and Benoît TROUILLET.

several studies implementing different models (Petri nets, constraint satisfaction problems, linear optimization problems), methods (analysis of Petri nets, constraint programming, linear optimization) and tools (heuristics, genetic algorithms, constraint solvers).

In this chapter, we focus on the problem of saturation of a railway junction. This problem consists of making a maximum number of trains go through a railway junction defined by a given set of lines that can cross each other and a number of safety rules.

To model this, we propose a representation based on timed Petri nets. These are widely exploited because they provide a simple, graphical model that takes the concepts of competition, parallelism and synchronization into account. For example, they are used in the study of communication networks, transportation systems or manufacturing.

Within the scope of this representation, an analogy can be made between train itineraries at a railway junction and the notion of an operating sequence in a job-shop. The saturation problem then boils down to a reachability problem, between the states of the Petri net. However, security rules affect the behavior of the system by prohibiting trains from being in neighboring blocks. Thus, a train, having crossed a block, cannot immediately return a resource and a deadlock phenomenon may occur, as trains mutually await a "block" resource before returning their own resource.

This kind of issue is a factor that restricts the application of some analysis algorithms, for example by inducing multiple false solutions. These are especially harmful for methods based on the construction and transversing of the reachability graph of a Petri net, as they can occur late in the Petri net execution sequence. Notably, this phenomenon is a strong limitation for classical scheduling algorithms, and was the subject of Damasceno's thesis [DAM 99]. We have chosen to address this because we believe that our approach is able to handle this problem in a natural manner.

Various methods have been suggested to address the reachability problem in Petri nets. Among them, we refer the interested reader to

Rauhamaa [RAU 90]. In this chapter, we detail two complementary approaches that allow us to solve the reachability problem in timed Petri nets.

The first approach directly builds on a general Petri net, constructing a mathematical model representing its behavior which can be solved using conventional branch and bound combinatorial optimization techniques. The second approach suggests modifying the network by eliminating conflicting resources to produce a simplified Petri net belonging to the class of event graphs, and then solve the timed reachability problem in this simplified net, using an algebraic technique.

The rest of the chapter is organized as follows: in section 5.2, we present issues regarding scheduling problems and we show how solving the railway junction saturation problem can be reduced to solving a scheduling problem in flexible systems. We formally define timed Petri nets in section 5.3, and show how they can represent the railway junction saturation problem in a natural manner. In sections 5.4 and 5.5, we develop the two approaches of proposed resolution, providing numerical experiments to validate the approach. We conclude by giving some promising research prospects.

5.2. Scheduling in a railway junction

5.2.1. *Classical scheduling*

In this section, we present the vocabulary necessary to define the scheduling problem. The vocabulary used corresponds to the description of a flexible manufacturing production system (which is the preferred field of application of many teams in the Nord Pas de Calais region), from which we borrow the concepts of machine, operation and range in their traditional meanings.

Scheduling a set of tasks means programming their execution by allocating the necessary resources and setting start dates. Scheduling problems [CHR 82] appear in all areas of economics: information technology (tasks are programs, resources are processors, memory, etc.), construction (project management), industry (workshop activities in production management and logistical problems), administration (timetables) and of course transport (road planning).

In a scheduling problem, two fundamental concepts are involved: tasks and resources. A resource is a technical or human means whose limited availability is not known *a priori*. A task is a basic job whose implementation requires a number of units of time (duration) and units of each resource. Dealing with production systems, resources correspond to machines and tasks are typically grouped together to form operating sequences. The scheduling problem then is to combine starting dates of tasks in such a way as to minimize the total production time (also called Cycle Time (CT) or makespan). In the context of transport problems, resources correspond to lines that can be used, and tasks correspond to the desired itineraries. The problem then is to allow a maximum number of vehicles to travel along these lines in a minimum time.

Scheduling problems are known to be complex and highly combinatorial [BEL 82]. It has been proven that project planning problems are polynomial and cyclic scheduling is NP-complete [CAR 88; SER 89]. The introduction of transformation resources (associated with different operations) makes the former a NP-hard problem in most cases, leaving the latter as NP-complete problems. The interested reader is referred to [DIA 93] for a complete bibliography on the subject.

5.2.2. *Flexible system scheduling*

We are interested in scheduling problems for flexible manufacturing systems (FMS). The FMS are characterized by the possibility of simultaneously producing different types of pieces, and the opportunity of using different procedures (the flexibilities) to produce the same type of piece.

The traditional approach in such cases is to proceed in two steps. In the first step, flexibilities are gradually resolved (e.g. through the assignment of tasks to machines, respecting routing ratios) to obtain a

Petri net where conflicts only correspond to the order of operations to perform. Solving the scheduling problem is then reduced to solving the corresponding reachability problem, which can be done through various techniques (heuristics, simulation, accurate methods for system subclasses, etc.).

5.2.3. *Dual Gantt diagram*

To represent a schedule, it helps to simultaneously visualize the scheduling of operations relative to machines, as well as the scheduling of operations within each operation sequence. The corresponding diagram is called a dual Gantt diagram. The advantage of such a representation is that it enables people to understand key scheduling parameters on a single diagram.

An example of a dual Gantt diagram is shown in Figure 5.1. The considered production is composed of three machines and three lines, with obvious notation described in Table 5.1.

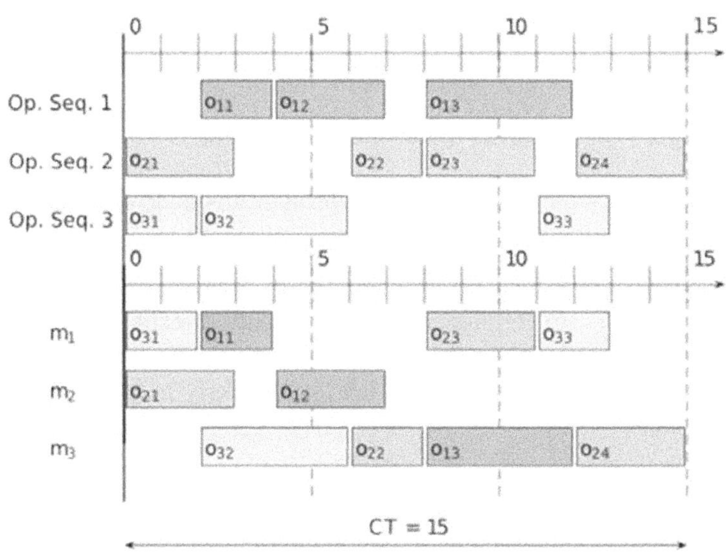

Figure 5.1. *Dual Gantt diagram*

Operating sequence 1	$(m_1,2)(m_2,3)(m_3,4)$
Operating sequence 2	$(m_2,3)(m_3,2)(m_1,3)(m_3,3)$
Operating sequence 3	$(m_1,2)(m_3,4)(m_1,2)$

Table 5.1. *Production schedule*

5.2.4. *The railway junction saturation problem*

A railway junction is a site featuring a large number of railways crossing, on which several types of train run. We can cite for example the Pierrefitte-Gonesse node, which was the subject of several studies [DEL 03; BEN 99; BOU 04a]. It is located about ten kilometers from Paris Nord station, France, and extends about sixteen kilometers. It represents a kind of crossroads of four directions (excluding RER (*Réseau Express Régional* – Regional Express Network) lines) where lines cross in many places. We distinguish between three main types of line in the Pierrefitte-Gonesse node:

– the mainline connecting Paris Nord and Chantilly;

– high speed lines connecting Paris Nord to Lille;

– lines connecting Chantilly to the Grande Ceinture.

Thus, the following types of train run through this node on a daily basis:

– High speed such as the TGV (to the north of France), Eurostar (to England) and Thalys (to Belgium) traveling in both directions. These usually use high-speed lines but occasionally can be diverted to the mainline tracks. This type of traffic is heavy and likely to increase in the years to come.

– The "classic" passenger trains between Paris Nord and Chantilly in both directions. Here again, train traffic is heavy. It should not, however, increase too much in the years to come.

– Freight trains between Chantilly and the Grande Ceinture. This type of traffic is more limited than the previous two but these trains are also much slower.

Moreover, other trains may occasionally use this node:

– commuter trains diverted via the mainline (RER station Pierrefitte-Gonesse is located in the area of study);

– special or exceptional trains whose routes in the node are diverse. The circulation of a towing vehicle is an example of this type of traffic.

It is therefore clear that this heavy traffic can cause many disruptions, notably freight trains that cut across all other lines whilst traveling at a much slower speed than other trains. In addition, the large number of lines allowing traffic in both directions does not improve the situation. Traffic optimization in such a node is far from easy.

In addition, special safety rules govern railway traffic which further complicates the problem. The railway junction is divided into areas called blocks, as shown in Figure 5.2. Each block represents the distance required for a train to come to a complete stop, and is the location of a traffic managing rule.

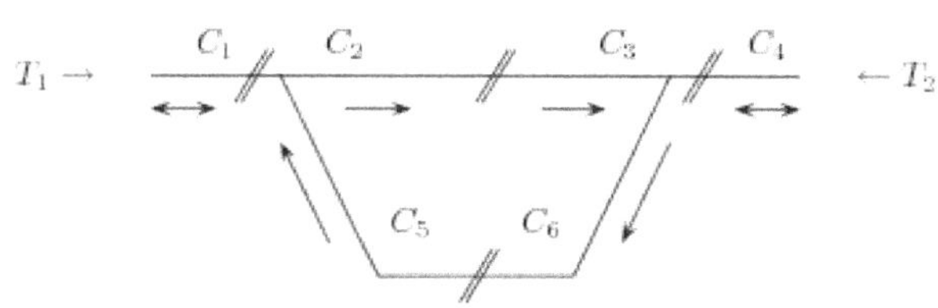

Figure 5.2. *Example of a railway junction*

The protection mechanism used to avoid common catching-up and nose-to-nose problems is to associate one resource C_i per block. Before committing to travel through block C_i at its nominal speed, train *Tk* reserves the block(s) located downstream of the block to travel through, which ensures that there will always be at least one empty block between two trains. The resource is freed when the train leaves the corresponding block.

This operation leads to the existence of deadlocks in the accessible states of the system, due to the consumption of shared resources in a particular order. For example, before moving into block C_1, the first train will have reserved blocks C_2 and C_5. Similarly, the second train engaged in C_4 will have reserved blocks C_3 and C_6. To move into C_6, and return the resources it has (including C_3), it will then have to reserve block C_5, currently assigned to the first train, which will be freed in exchange for the C_3 resource, and so on.

We have chosen to focus on this type of issue because we believe that our approaches are able to handle this problem in a natural manner. Our approaches are based on modeling the saturation problem in the form of a reachability problem in timed Petri nets. We present this modeling approach in the next section.

5.3. Petri nets for scheduling

5.3.1. *Place/Transition Petri net*

5.3.1.1. *Bipartite graph*

A Place/Transition Petri net (P/T PN) [DIA 93] (R = ($\mathbb{P}$, W), m0) is a directed bipartite graph where:

– $\mathbb{P} = \{p_1,...,p_M\}$ and $\mathbb{T} = \{t_1,...,t_N\}$ are two finite sets of nodes, called places and transitions respectively, with $|\mathbb{P}| = M$ and $|\mathbb{T}| = N$;

– W: $\mathbb{P} \times \mathbb{T} \cup \mathbb{T} \times \mathbb{P} \rightarrow \mathbb{N}$ represents the weights of arcs connecting the nodes;

– m_0: $\mathbb{P} \rightarrow \mathbb{N}$ associates with each place $p \in \mathbb{P}$ a whole $m_0\,(p)$ called marking p.

The terms "upstream" and "downstream" will be used to respectively denote the predecessors and successors of a node in the network.

5.3.1.2. *Graphical representation*

In the graphical representation of a Petri net, places are represented as circles and transitions as rectangles. Tokens are represented by filled discs inside places or simply by an integer labeling the place. Figure 5.3 shows an example of a P/T PN, where $m_0 = (1,0,0,0,1,0,1,1)^T$.

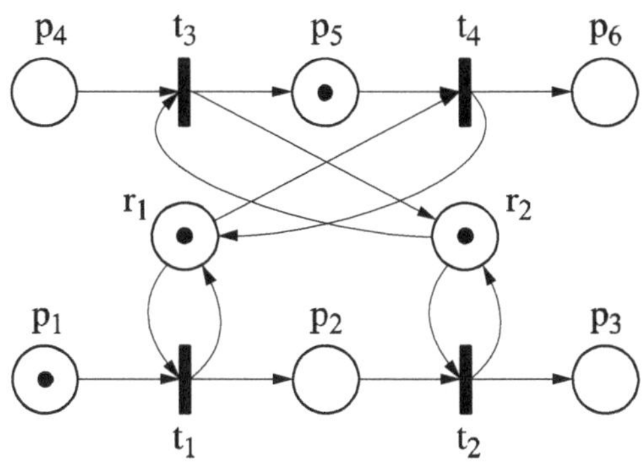

Figure 5.3. *Example of a P/T PN*

5.3.1.3. *Matrix representation*

Incidence matrices C^-, $C^+ \in \mathbb{N}^{M\times N}$ and $C \in \mathbb{Z}^{M\times N}$ are defined by:

$$\begin{matrix} \forall i \in [\![1, M]\!] \\ \forall j \in [\![1, N]\!] \end{matrix} \begin{cases} C^-(i,j) = k \Leftrightarrow W(p_i, t_j) = k \\ C^+(i,j) = k \Leftrightarrow W(t_j, p_i) = k \\ C(i,j) = C^+(p_i, t_j) - C^-(p_i, t_j) \end{cases}$$

Incidence matrices corresponding to the network in Figure 5.3 are depicted in Figure 5.4.

$$\underset{C^-}{\begin{bmatrix} 1 & 0 & 0 & 1 \\ 0 & 1 & 0 & 0 \\ 0 & 0 & 0 & 0 \\ 0 & 0 & 0 & 0 \\ 0 & 0 & 0 & 1 \\ 0 & 0 & 0 & 0 \\ 1 & 0 & 1 & 1 \\ 0 & 1 & 0 & 0 \end{bmatrix}} \quad \underset{C^+}{\begin{bmatrix} 0 & 0 & 0 & 0 \\ 1 & 0 & 0 & 0 \\ 0 & 1 & 0 & 0 \\ 0 & 0 & 0 & 0 \\ 0 & 0 & 1 & 0 \\ 0 & 0 & 0 & 1 \\ 1 & 0 & 0 & 1 \\ 0 & 1 & 1 & 0 \end{bmatrix}} \quad \underset{C}{\begin{bmatrix} -1 & 0 & 0 & 0 \\ 1 & -1 & 0 & 0 \\ 0 & 1 & 0 & 0 \\ 0 & 0 & -1 & 0 \\ 0 & 0 & 1 & -1 \\ 0 & 0 & 0 & 1 \\ 0 & 0 & 0 & 0 \\ 0 & 0 & 0 & 0 \end{bmatrix}}$$

Figure 5.4. *Incidence matrices of the network in Figure 5.3*

Below, we will use linear algebra notation to make the information more concise:

– Vector $\overrightarrow{e_{t_k}}$ denotes the Parikh (or characteristic) vector associated with transition t_k, whose kth component equals 1 and the others are zero.

– Vector $\overrightarrow{m_0}$ denotes the marking vector, defined by $\overrightarrow{m_0} = (m(p1), m(p2), \dots, m(pM))^T \in \mathbb{N}^M$.

5.3.1.4. *Behavior*

A transition t_k is said to be fireable for a given marking m_0 (denoted $m_0[t_k\rangle$) if the places upstream of the transition contain enough tokens:

$$m_0[tk\rangle \begin{cases} \Leftrightarrow \forall p \in \mathbb{P},\ m_0(p) \geq W(p, t_k) \\ \Leftrightarrow \qquad \overrightarrow{m_0} \geq C^-.\overrightarrow{e_{t_k}} \end{cases}$$

Firing of t_k then produces a new marking m_1 from marking m_0 (denoted $m_0[t_k\rangle m_1$):

$$m_0[tk\rangle m1$$

$$\begin{aligned} &\Leftrightarrow m_0[tk\rangle \wedge \\ &\qquad \forall p \in \mathbb{P},\ m_1(p) = m_0(p) - W(p, t_k) + W(t_k, p) \\ &\Leftrightarrow \quad m_0[tk\rangle \wedge \overrightarrow{m_1} = \overrightarrow{m_0} + C \ \ .\overrightarrow{e_{t_k}} \end{aligned}$$

5.3.1.5. *State equation*

The above equations generalize the firing of fireable transition sequences. For this, we define the characteristic vector $\vec{\sigma}$ associated with a sequence of r transitions of $\mathbb{T}$: $\sigma = t_{\sigma_1} t_{\sigma_2} \dots t_{\sigma_r}$, as the vector of $\mathbb{N}^N$ for which the jth component is equal to the number of occurrences of the t_j in σ:

$$\vec{\sigma} = \sum_{j=1}^{r} \overrightarrow{e_{\sigma_j}}$$

By using the characteristic vector and previous algebraic notation, we obtain a classical property of Petri nets:

$$m_0[\sigma\rangle m_f \Rightarrow \overrightarrow{M_f} = \overrightarrow{M_0} + C.\vec{\sigma}.$$

This equation is known as the state equation or fundamental equation of Petri nets. The reverse implication (which is not true in most cases) allows us to illustrate the famous Petri net reachability problem, which is expressed as follows:

> "Given an initial marking m_0 and a final marking m_f, find a fireability sequence σ such that $m_0[\sigma\rangle m_f$."

5.3.2. *T-timed Petri nets*

T-timed Petri nets (TPNs) were introduced by Ramchandani [RAM 74]. The following description is from Chrétienne [CHR 84a].

– TPNs correspond to P/T PN where a time $d(t)$ is associated with each transition t: $\forall t \in \mathbb{T}$, $d(t) \in \mathbb{N}$. Note that we could more generally consider Petri nets in which times are rational valued. However, after reducing them to the same denominator and matching up the numerators, it appears as if they were integers. The above definition does not forbid the existence of transitions of null time, which are referred to as instantaneous transitions. However, to simplify the study, we will restrict ourselves in the following to the case of timed Petri nets without instantaneous transitions, which is not too restrictive in real case studies. In the end, we obtain: $\forall t \in \mathbb{T}$, $d(t) \in \mathbb{N}^*$.

– A TPN is graphically represented in the same way as P/T PN, for which *labeling* is added to transitions. An example of a TPN is shown in Figure 5.5. We obtain: $d(t_1) = 3$, $d(t_2) = 4$, $d(t_3) = 5$, $d(t_4) = 2$.

– Firing durations associated with transitions of a TPN modify the validity conditions of markings. All this occurs as if tokens disappear when the transition is fired, then reappear after a period corresponding to the duration of the transition firing. Markings therefore evolve according to the occurrences of clock signals external to Petri nets.

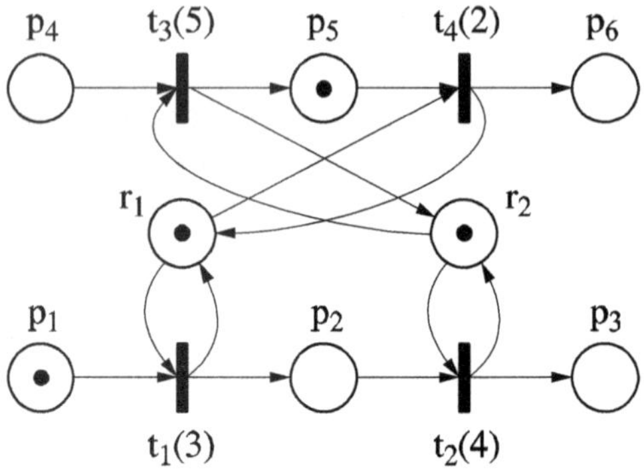

Figure 5.5. *Example of a TPN*

For example, let us consider the TPN in Figure 5.5. If we fire transition t_1 of duration 3 at date 1, then transition t_4 of duration 2 at date 5, the evolution of markings over time is given in Table 5.2.

	Date	Marking
Initial date	0	$(1,0,0,0,1,0,1,1)^T$
Firing of t_1 →	1	$(0,0,0,0,1,0,0,1)^T$
	2	$(0,0,0,0,1,0,0,1)^T$
	3	$(0,0,0,0,1,0,0,1)^T$
End of firing of t_1	4	$(0,1,0,0,1,0,1,1)^T$
Firing of t_4 →	5	$(0,1,0,0,0,0,0,1)^T$
	6	$(0,1,0,0,0,0,0,1)^T$
End of firing of t_4	7	$(0,1,0,0,0,1,1,1)^T$
	8	$(0,1,0,0,0,1,1,1)^T$

Table 5.2. *TPN firings*

Note that we could already fire transition t_4 at date 4 since resource r_1 was returned at the end of firing t_1. However, this transition was not fireable at date 3, since firing the transition t_1 had yet to be completed.

Thus, the starting and ending dates of transition firings play a fundamental role in the behavior of a timed Petri net. It is therefore necessary to express the fireability conditions at each date we want to fire a transition. To do this, it is necessary to introduce the concept of controlled execution that corresponds to the history of fired transitions in the Petri net, allowing us to postulate on the state reached by the Petri net at a given date.

5.3.3. *Controlled executions*

The fundamental concept that governs the behavior of a TPN is *controlled execution*, which associates with each transition the start dates of successive firings. We denote by $(u_k^t) = u_1^t, \ldots, u_{K_t}^t \in \mathbb{N}$ the increasing sequence of firing instances of transition t. The controlled execution is then the family of firing sequences $(u_k^t)_{t \in \mathbb{T}, k \in [\![1, K_t]\!]}$.

For consistency reasons, it is common to assume that the semantics of transition firing forbids re-entrance, meaning that it is not possible to fire the same transition several times, and that we cannot once again fire a transition that has not finished being fired. In this way, we can associate a single duration, called residual, to each transition, without it being possible to confuse between multiple concurrent activations of the same transition.

The state of a TPN subject to a controlled execution can be defined at any instant $v \in \mathbb{N}$ by the data from its marking vector $\vec{e}_m(v) \in \mathbb{Z}^{\mathrm{M}}$ and its residual duration vector $\vec{e}_r(v) \in \mathbb{N}^{\mathrm{N}}$.

To comply with the underlying P/T PN semantics, a controlled execution $(u_k^t)_{t \in \mathbb{T}, k \in [\![1, K_t]\!]}$ is said to be feasible if and only if, at any time, the marking vector remains a positive or null. To conclude, it is obviously sufficient to check this property only at firing dates.

5.3.4. *Reachability problems in TPNs*

With the above notations, the timed Petri net reachability problem consists of finding a feasible controlled execution to move from an initial state $(\overrightarrow{e_{m0}}, \overrightarrow{e_{r0}})$ at date v_0, to a final state $(\overrightarrow{e_{mf}}, \overrightarrow{e_{rf}})$ at date v_f.

As mentioned in the introduction, it is very simple to make a parallel between a scheduling problem and a TPN reachability problem. For example, consider the Petri net in Figure 5.6. We can clearly see that the reachability problem between markings

$$m_0 = \{p_1, p_2, p_3, m_1, m_2, m_3\}$$

and

$$m_f = \{p_4, p_5, p_6, m_1, m_2, m_3\}$$

is the same as the scheduling problem as presented in Table 5.1.

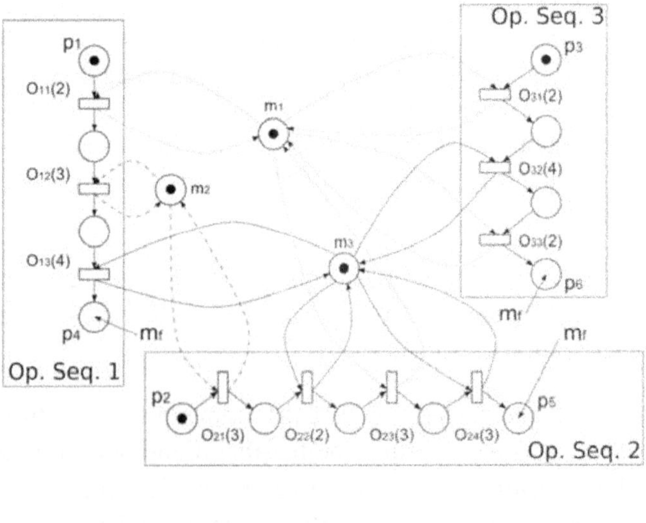

Figure 5.6. *TPN modeling the production in Table 5.1*

5.3.5. *Modeling of a railway junction with Petri nets*

As with any scheduling problem, the behavior of trains at a railway junction is represented in a natural way in Petri nets. "Resource" places are the blocks, tokens denote the occupation of these blocks by trains and timed transitions correspond to their itineraries. Thus, the Petri net modeling the railway junction illustrated in Figure 5.2 is shown in Figure 5.7.

Each place C_i represents a block of the same name, and operates as a shared resource, which evokes the notion of flexibility, as introduced above (with the difference that the execution of a "task" may require several "machines"). Before committing to travel through block C_i at the nominal speed corresponding to the timed transition, labeled "T_k in C_i", train T_k reserve(s) the block(s) located downstream of the block through which to travel; by consuming tokens located in associated resource places. The resource is released when the train leaves the corresponding block. When they have no explicit indication of time, transitions have a unitary duration.

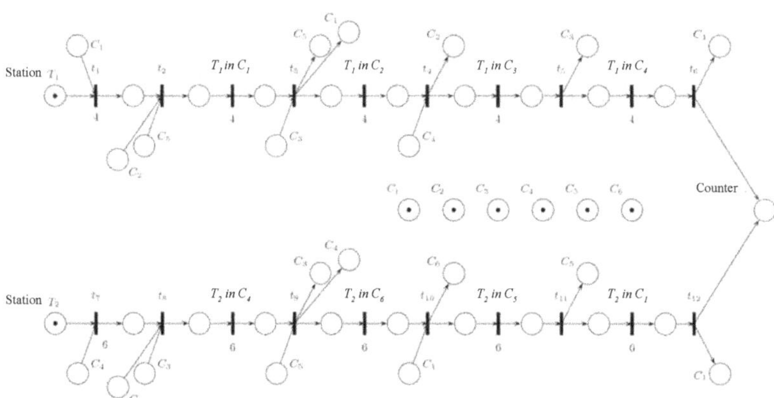

Figure 5.7. *TPN modeling the railway junction in Figure 5.2*

This example demonstrates a classical phenomenon that increases the complexity of studies to be conducted: the existence of "*dead markings*" in reachable states of the Petri net, corresponding to system deadlocks due to the consumption of shared resources in a particular order. For example, if we consider the firing of t_1, t_2, t_7 and t_8, all resources associated with blocks are consumed. The system is blocked because in order to return the resources they control, t_3 and t_9 transitions must first have an additional resource.

This kind of issue is not strictly speaking a "cause" of combinatorial explosion, but it is a factor that restricts the application of some analysis algorithms, for example by inducing multiple false solutions. These are especially harmful for methods based on the construction and traversal of the reachability graph which can occur late in the network execution sequence. This phenomenon has particularly strong limitations for classical scheduling algorithms [DAM 99].

5.3.6. *Approaches to solving the timed reachability problem*

Several studies have focused on the search for the scheduling solutions for the TPN reachability problem, either limited to a particular subclass of TPNs as timed event graphs, or using dedicated heuristics (see [RIC 97] for a complete bibliography).

Since the firing of a timed transition can take place as soon as it is fireable and as late as one wants, there may exist an infinite number of accessible markings (based on residual duration) from a given state, and it is not possible to build a reachability graph. One approach is to consider timed Petri nets as a subclass of temporal Petri nets, in order to benefit from enumerative approaches (state class graph) as proposed by [BER 91].

Another approach is to look at the earliest firing semantics (a transition is fired as soon as it is fireable). In this case, it is possible to proceed to enumerative and structural analysis [DAV 92]. However, it should be noted that earliest firing semantics does not guarantee scheduling optimality.

For example, let us consider the timed Petri net in Figure 5.8. Once the transitions t_1 and t_4 are fired in parallel, the earliest operation is then to cross t_5, which leads to the firing sequence $\sigma_1 = t_1 +$

$t_{4(v1=0)}$, $t_{5(v2=1)}, t_2 + t_{6(v3=11)}$, $t_{3(v4=12)}$ for a duration of 22 *t.u.* However, optimal operation is obtained by delaying the firing of t_5 to make sequence $\sigma_2 = t_1 + t_{4(v1=0)}$, $t_{2(v2=2)}, t_3 + t_{5(v3=3)}$, $t_{6(v4=13)}$ last 14 *t.u.* Thus, to achieve the optimal plan, it is necessary to solve the conflict between transitions t_2 and t_5 without respecting the earliest firing policy.

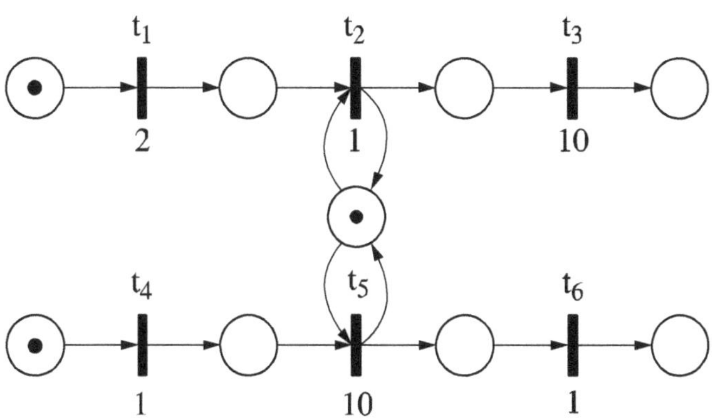

Figure 5.8. *TPN and earliest firing semantics*

The semantics of the earliest firing has been intensively studied for the particular class of timed event graphs, using (max,+) algebra [BAC 92]. Since the structure is conflict-free, it is possible to obtain linear equations corresponding to the complete system behavior and to ensure the optimality of the earliest firing semantics [CHR 84b].

In the next two sections, we propose to solve the reachability problem in general Petri nets using accurate methods. The first section consists of producing an integer linear programming model characterizing the behavior of a general TPN.

5.4. Incremental model for TPNs

In this section, we present an approach for the resolution of reachability problems in TPNs, based on mathematical programming techniques. For this, we propose to use a mathematical model to represent the complete behavior of a TPN. Unlike conventional work using TPNs, we do not make any hypotheses on the class of Petri nets studied, which makes our model as general as possible. The advantage of our approach is to allow *choice structures* in the Petri net, and thus to be able to directly study a non-deterministic model, while guaranteeing the accuracy of the approach. Of course, the induced complexity is greater however it is a first approach in this direction which should be examined in light of the progress made by modern solvers. In addition, we can use the wide range of improvements (modular decomposition, cuts, relaxations, etc.) offered by this approach.

Our model is based on an incremental constraint propagation with an increasing number of transition firings of a controlled execution.

Let $\mathbb{V} = \{v_1,\dots,v_{V-1},v_V = v_f\}$ with $|\mathbb{V}| = V \in \mathbb{N}^*$ the set of firing dates $v_k \in \mathbb{N}$ of the considered controlled execution. Let $\vec{\sigma}(v_k) \in \mathbb{N}^N$ be the Parikh vector fired transitions at date $v_k \in \mathbb{V}$.

We will express on every firing date $v_k \in \mathbb{V}$ the changes induced on vectors $\overrightarrow{e_m}(vk)$ and $\overrightarrow{e_r}(vk)$ from the firing of $\vec{\sigma}(v_k)$, and firing constraints $\vec{\sigma}(v_k)$.

To be fireable, transitions $\vec{\sigma}(v_k)$ must not *already be firing at date* v_k, following a firing triggered at an earlier time. By denoting $\Delta_{v_{k-1}}$ the period between the dates $k-1$ and k: $\forall k \in [\![1,V]\!]$, $\Delta_{v_{k-1}} = v_k - v_{k-1}$, this constraint of *non-re-entrance* can be expressed as follows:

$$\forall k \in [\![1,V]\!], \forall t \in \mathbb{T},$$

$$\overrightarrow{e_r}(v_{k-1})|_t > \Delta_{v_{k-1}} \quad \Rightarrow \vec{\sigma}(v_k)|_t = 0 \qquad [5.1]$$

In addition, there must be sufficient tokens in the Petri net places at date v_k to cross $\vec{\sigma}(v_k)$. This is the case if, after consuming the tokens needed for firing $\vec{\sigma}(v_k)$, no place has a negative marking, and is expressed by the constraint:

$$\forall\, k \in [\![1, \mathrm{V}]\!] \;\; \overrightarrow{e_m}(v_k) \geq \vec{0} \qquad [5.2]$$

To search for controlled executions leading from the initial state to the final state, it is sufficient to impose the constraints:

$$\overrightarrow{e_m}(v_0) = \overrightarrow{e_{m0}} \qquad [5.3]$$

$$\overrightarrow{e_r}(v_0) = \overrightarrow{e_{r0}} \qquad [5.4]$$

$$\overrightarrow{e_m}(v_\mathrm{V}) = \overrightarrow{e_m}\left(\sum_{k \in [\![0,\mathrm{V}-1]\!]} \Delta_{v_k}\right) = \overrightarrow{e_{mf}} \qquad [5.5]$$

$$\overrightarrow{e_r}(v_\mathrm{V}) = \overrightarrow{e_r}\left(\sum_{k \in [\![0,\mathrm{V}-1]\!]} \Delta_{v_k}\right) = \overrightarrow{e_{rf}} \qquad [5.6]$$

To express the values of updated state vectors, we use the following conventions. If $\vec{x}$ is a vector with values in $\mathbb{Z}^k$, we will denote by:

– $\vec{x}^+ \in \mathbb{N}^k$, the vector of its positive components, such that $\forall c \in [\![1, k]\!]$, $\vec{x}^+(c) = \vec{x}(c)$ if $\vec{x}(c) > 0$ and 0 otherwise;

– $\vec{x}^+ \in \mathbb{B}^k$, a vector representing its sign, such that: $\forall c \in [\![1, k]\!]$, $\vec{x}^s(c) = \vec{x}(c)$ if $\vec{x}(c) \leq 0$ and $\vec{x}^s(c)$=1 otherwise.

We will see later that through a number of intermediate variables and constraints, it is possible to formulate the values of these characteristic vectors linearly.

The previous notations allow us to affirm that:

$$\forall k \in [\![1, V]\!],$$

$$\begin{aligned}\overrightarrow{e_m}(v_k) &= \overrightarrow{e_m}(v_{k-1}) - C^- \vec{\sigma}(v_k) \\ &+ C^+ \left(\overrightarrow{e_r}(v_{k-1})^s - \left(\overrightarrow{e_r}(v_{k-1}) - \Delta_{v_{k-1}} \overrightarrow{Id} \right)^s \right)\end{aligned}$$

[5.7]

$$\begin{aligned}\overrightarrow{e_r}(v_k) &= \left(\overrightarrow{e_r}(v_{k-1}) - \Delta_{v_{k-1}} \overrightarrow{Id} \right)^+ \\ &+ \sum_{t \in \mathbb{T}} d(t) . \vec{\sigma}(v_k)|_t . \overrightarrow{e_t}\end{aligned}$$

[5.8]

Of course, the same equations remain valid if no transition is fired at date v_k (i.e. when $\vec{\sigma}(v_k) = \vec{0}$), and are then used to calculate the instantaneous state at that date.

Constraint [5.1] that prohibits re-entrant firings can also be translated using the above notation, introducing the inequality:

$$\forall k \in [\![1, V]\!], \vec{\sigma}(v_k) \leq \overrightarrow{1_N} - \left(\overrightarrow{e_r}(v_{k-1}) - \Delta_{v_{k-1}} . \overrightarrow{1_N} \right)^s \qquad [5.9]$$

Indeed, if transition t is still active at the time v_k after the firing that was previously activated, then: $\left(\overrightarrow{e_r}(v_{k-1}) - \Delta_{v_{k-1}} . \overrightarrow{1_N} \right)\Big|_t > 0$ and therefore: $\left(\overrightarrow{e_r}(v_{k-1}) - \Delta_{v_{k-1}} . \overrightarrow{1_N} \right)^s\Big|_t = 1$. Equation [5.9] then gives: $\vec{\sigma}(v_k)|_t \leq 0$, which prohibits t from being fired in $\vec{\sigma}(v_k)$.

Conversely, if t was not active at date v_k, we would have: $\left(\overrightarrow{e_r}(v_{k-1}) - \Delta_{v_{k-1}} \overrightarrow{1_N} \right)^s\Big|_t = 0$. Thus, equation [5.9] gives: $\vec{\sigma}(v_k)|_t \leq 1$, which is still a valid inequality since $\vec{\sigma} \in \{0, 1\}^N$.

A complete demonstration of the validity of the above equations regarding controlled executions would be outside the scope of this chapter. Thus we propose to illustrate this calculation around an example in order to give main ideas. Consider the TPN in Figure 5.5, given with its initial state at date $v_0 = 0$:

$$\begin{cases} \overrightarrow{e_m}(v_0) = (1,0,0,0,1,0,1,1)^{\mathrm{T}} \\ \overrightarrow{e_r}(v_0) = (0,0,0,0)^{\mathrm{T}} \end{cases}$$

Assume that we fire t_1 at date $v_1 = 1$, then simultaneously t_2 and t_4 at date $v_2 = 4$, and we want to calculate the final state at date $v_3 = 6$. The calculation steps are shown in Tables 5.3 and 5.4. It is easy to understand the physical meaning of each of the components of the equations:

– The quantity $\left(\overrightarrow{e_r}(v_{k-1}) - \Delta_{v_{k-1}}\overrightarrow{Id}\right)^+$ is the updated residual duration vector at date v_k from its value $\overrightarrow{e_r}(v_{k-1})$ at date v_{k-1}. The "+" operator allows us to take only the positive values into account.

– The quantity $\sum_{t\in\mathbb{T}} d(t).\vec{\sigma}(v_k)|_t.\overrightarrow{e_t}$ represents new residual durations from the execution of the sequence of transitions $\vec{\sigma}(v_k)$ at date v_k.

– The quantity $C^-.\vec{\sigma}(v_k)$ represents the number of tokens removed from places upstream of the sequence of transitions $\vec{\sigma}(v_k)$, whose firing *starts* at date v_k.

– Finally, the quantity $\overrightarrow{e_r}(v_{k-1})^s - \left(\overrightarrow{e_r}(v_{k-1}) - \Delta_{v_{k-1}}\overrightarrow{Id}\right)^s$ is the characteristic transitions' vector whose execution *ends* between v_{k-1} and v_k. This expression is based on a comparison between the characteristic vector of transitions that were being fired at date v_{k-1} (vector $\overrightarrow{e_r}(v_{k-1})^s$), and that of transitions that are active at date v_k (vector $\left(\overrightarrow{e_r}(v_{k-1}) - \Delta_{v_{k-1}}\overrightarrow{Id}\right)^s$).

Equations [5.7] and [5.8] allow us to write an integer linear programming model, able to represent the complete behavior of a TPN; the only assumption made is the absence of instantaneous transitions (see equation [5.8] that does not take instantaneous firings for which $d(t) = 0$ into account).

In the next section, we present a linear formulation to express the action of operators defined above.

5.4.1. *Formulation operators "+" and "s"*

Operators "+" and "s" introduced in the previous section operate on vectors of variables. However, they act uniformly on each component of the input vector, and may be naturally defined by the application of an equivalent operator acting on scalar values. In this section, we are interested in calculating the values X^s and X^+ associated with an integer $X \in \mathbb{Z}$ such that:

$$\begin{cases} X > 0 & \Rightarrow & X^s = 1 \text{ and } X^+ = X \\ X \leq 0 & \Rightarrow & X^s = 0 \text{ and } X^+ = 0 \end{cases}$$

To do this, we propose to introduce two variables and five additional equations. We justify our approach by the following statement:

PROPOSITION 5.1.– (Discrimination variables)

Let $X \in \mathbb{Z}$ and $B \in \mathbb{N}$ "sufficiently large".

Let $\alpha \in \{0,1\}$ and $\beta \in \mathbb{N}$ such that:

$$1 + (\alpha - 1) \cdot B \;)\; X \;)\; (t \cdot B \qquad [5.10]$$

$$X \;.\; \beta \qquad [5.11]$$

$$\beta \leq \alpha \cdot B \qquad [5.12]$$

$$\beta \leq B \cdot (1 - \alpha) + X \qquad [5.13]$$

We obtain:

$$\begin{cases} X > 0 & \Rightarrow & \alpha = 1 \text{ and } \beta = X \\ X \leq 0 & \Rightarrow & \alpha = 0 \text{ and } \beta = 0 \end{cases} \qquad [5.14]$$

$v_0 = 0$		$\Delta v_0 = 1, v_1 = 1$			$\Delta v_1 = 3, v_2 = 4$			$\Delta v_2 = 2, v_3 = 6$		
places	$\overrightarrow{e_m}(v_0)$	$C^-.\vec{\sigma}(v_1)$	$C^+.\left(\overrightarrow{e_r}(v_0)^s - \left(\overrightarrow{e_r}(v_0) - \Delta_{v_0}\overrightarrow{Id}\right)^s\right)$	$\overrightarrow{e_m}(v_0 + \Delta v_0) = \overrightarrow{e_m}(v_1)$	$C^-.\vec{\sigma}(v_2)$	$C^+\left(\overrightarrow{e_r}(v_1)^s - \left(\overrightarrow{e_r}(v_1) - \Delta_{v_1}\overrightarrow{Id}\right)^s\right)$	$\overrightarrow{e_m}(v_1 + \Delta v_1) = \overrightarrow{e_m}(v_2)$	$C^-.\vec{\sigma}(v_3)$	$C^+.\left(\overrightarrow{e_r}(v_2)^s - \left(\overrightarrow{e_r}(v_2) - \Delta_{v_2}\overrightarrow{Id}\right)^s\right)$	$\overrightarrow{e_m}(v_2 + \Delta v_2) = \overrightarrow{e_m}(v_3)$
*p*1	1	1	0	0	0	0	0	0	0	0
*p*2	0	0	0	0	1	1	0	0	0	0
*p*3	0	0	0	0	0	0	0	0	0	0
*p*3	0	0	0	0	0	0	0	0	0	0
*p*4	1	0	0	1	1	0	0	0	0	0
*p*5	0	0	0	0	0	0	0	0	1	1
*r*1	1	1	0	0	1	1	0	0	1	1
*r*2	1	1	0	1	1	0	0	0	0	0

Table 5.3. *Intermediate calculations - markings*

We will present an informal proof of the previous mechanism. Suppose that X is strictly positive. The inequality on the right in equation [5.10] then implies that $1 \leq X \leq \alpha \cdot B$, or $\alpha \geq \frac{1}{B} > 0$. Therefore, we obtain $\alpha = 1$. Equation [5.13] then requires $\beta \leq X$, and equation [5.11] induces $\beta = X$. Equation [5.12] is satisfied if B is large enough.

Conversely, if X is negative or zero, the left side of equation [5.10] then implies that $1 + (\alpha - 1) \cdot B \leq 0$, which requires that $\alpha \leq 1 - \frac{1}{B} < 1$. Therefore, we get $\alpha = 0$. Equation [5.12] then requires $\beta \leq 0$ or $\beta = 0$. Equations [5.11] and [5.13] are obviously verified if B is large enough.

	transitions	t_1	t_2	t_3	t_4
$v_0 = 0$	$d(t)$	3	4	5	2
	$\overrightarrow{e_r}(v_0)$	0	0	0	0
$\Delta v_0 = 1,$ $v_1 = 1$	$\vec{\sigma}(v_1)$	1	0	0	0
	$\overrightarrow{e_r}(v_0)^s$	0	0	0	0
	$\overrightarrow{e_r}(v_0) - \Delta_{v_0}\overrightarrow{Id}$	-1	-1	-1	-1
	$\left(\overrightarrow{e_r}(v_0) - \Delta_{v_0}\overrightarrow{Id}\right)^s$	0	0	0	0
	$\left(\overrightarrow{e_r}(v_0) - \Delta_{v_0}\overrightarrow{Id}\right)^+$	0	0	0	0
	$\sum_{t\in\mathbb{T}} d(t).\vec{\sigma}(v_1)\|_t.\overrightarrow{e_t}$	3	0	0	0
	$\overrightarrow{e_r}(v_0)^s - \left(\overrightarrow{e_r}(v_0) - \Delta_{v_0}\overrightarrow{Id}\right)^s$	0	0	0	0
	$\overrightarrow{e_r}(v_0 + \Delta v_0) = \overrightarrow{e_r}(v_1)$	3	0	0	0
$\Delta v_1 = 3,$ $v_2 = 4$	$\vec{\sigma}(v_2)$	0	1	0	1
	$\overrightarrow{e_r}(v_1)^s$	1	0	0	0
	$\overrightarrow{e_r}(v_1) - \Delta_{v_1}\overrightarrow{Id}$	0	-3	-3	-3
	$\left(\overrightarrow{e_r}(v_1) - \Delta_{v_1}\overrightarrow{Id}\right)^s$	0	0	0	0
	$\left(\overrightarrow{e_r}(v_1) - \Delta_{v_1}\overrightarrow{Id}\right)^+$	0	0	0	0
	$\sum_{t\in\mathbb{T}} d(t).\vec{\sigma}(v_2)\|_t.\overrightarrow{e_t}$	0	4	0	2
	$\overrightarrow{e_r}(v_1)^s - \left(\overrightarrow{e_r}(v_1) - \Delta_{v_1}\overrightarrow{Id}\right)^s$	1	0	0	0
	$\overrightarrow{e_r}(v_1 + \Delta v_1) = \overrightarrow{e_r}(v_2)$	0	4	0	2
$\Delta v_2 = 2,$ $v_3 = 6$	$\vec{\sigma}(v_3)$	0	0	0	0
	$\overrightarrow{e_r}(v_2)^s$	0	1	0	1
	$\overrightarrow{e_r}(v_2) - \Delta_{v_2}\overrightarrow{Id}$	-2	2	-2	0
	$\left(\overrightarrow{e_r}(v_2) - \Delta_{v_2}\overrightarrow{Id}\right)^s$	0	1	0	0
	$\left(\overrightarrow{e_r}(v_2) - \Delta_{v_2}\overrightarrow{Id}\right)^+$	0	2	0	0
	$\sum_{t\in\mathbb{T}} d(t).\vec{\sigma}(v_3)\|_t.\overrightarrow{e_t}$	0	0	0	0
	$\overrightarrow{e_r}(v_2)^s - \left(\overrightarrow{e_r}(v_2) - \Delta_{v_2}\overrightarrow{Id}\right)^s$	0	0	0	1
	$\overrightarrow{e_r}(v_2 + \Delta v_2) = \overrightarrow{e_r}(v_3)$	0	2	0	0

Table 5.4. *Intermediate calculations: residual durations*

Setting B to be "sufficiently large" determines the validity of the inequations [5.10]. Indeed, since the parameter α is binary, B must obviously verify $1 - B \leq X \leq B$ for any possible X. The value of B must be selected to be strictly greater than the maximum possible values of X.

During the operational implementation of such constraints, we ensure dimensioning of the constant B so as not to exceed the computer representation and used resolution libraries' capacity, so it is always possible to differentiate $\frac{1}{B}$ from 0.

With the addition of a number of variables and constraints, we saw that it was possible to propose a linear expression of the operators used in equations [5.2] to [5.9]. It is therefore possible to propose a general linear mathematical model to solve the reachability problem in timed Petri nets.

5.4.2. *Integer Mathematical Models*

Notations that will be used to express the mathematical model are given below:

– the total number of firing dates is denoted by $V \in \mathbb{N}+$;

– the initial marking m_0 is given at date 0 without loss of generality. A first crossing of step σ_1 provides access to marking m_1 from m_0;

– the kth step firing σ_k is denoted by its characteristic vector $\overrightarrow{\sigma_k} \in \{0,1\}^N$;

– the state reached after the kth firing is defined by its marking vector $\overrightarrow{e_{mk}} \in \mathbb{N}^M$ and residual duration vector $\overrightarrow{e_{rk}} \in \mathbb{N}^N$;

– the dates of the firing of steps in the Petri net are denoted by $\overrightarrow{v_k} \in \mathbb{R}^+$. To simplify the formulation of equations, we also use the elapsed time between two successive firings: $\Delta_{v_k} = v_{k+1} - v_k \in \mathbb{N}$ (possibly with $v_k = v_{k+1}$).

To facilitate reading of the complete mathematical model, the previous notations are listed in Table 5.5.

e_{m_0}	$\longrightarrow$	$e_{m_1} \cdots e_{m_k}$	$\longrightarrow$	$e_{m_{k+1}} \cdots e_{m_{V-1}}$	$\longrightarrow$	e_{m_V}
	σ_1		σ_{k+1}		σ_V	
$v_0 = 0$	$v_1 = v_0 + \Delta_{v_0}$		v_{k+1}		$v_V = v_{V-1} + \Delta_{V-1}$	
e_{r_0}	$\longrightarrow$	$e_{r_1} \cdots e_{r_k}$	$\longrightarrow$	$e_{r_{k+1}} \cdots e_{r_{V-1}}$	$\longrightarrow$	e_{r_V}

Table 5.5. *Synoptic table of notations*

The complete model is shown in Figure 5.9. Equations [5.16] to [5.19] correspond to the conditions on the initial and final states. Inequalities [5.20] to [5.26] express constraints on the discrimination variables used to calculate the operators "+" and "s". Variables (a_k), (α_k) and (β_k) respectively denote the values of $(\overrightarrow{e_r}(v_k)^s, \left(\overrightarrow{e_r}(v_k) - \Delta_{v_k}\overrightarrow{Id}\right)^s$ and $\left(\overrightarrow{e_r}(v_k) - \Delta_{v_k}\overrightarrow{Id}\right)^+$ used in equations [5.7] and [5.8]. Equations [5.27] and [5.29] are the equations for updating intermediate states [5.7] and [5.8]. Finally, equations [5.30] to [5.36] define the dimension of variables used.

Let $(R, \overrightarrow{m_0}, \overrightarrow{r_0}, d)$ be a TPN with $\mathbb{P} = \{p_1,\dots,p_M\}$, $\mathbb{T} = \{t_1,\dots,t_N\}$. Let C^+ and $C^- \in \mathbb{N}^{M \times N}$ be its incidence matrix. Let $(\overrightarrow{m_f}, \overrightarrow{r_f})$ be the final state defining the reachability problem. Let $V \in \mathbb{N}$. The ineger linear programming models $IP(V)$ is defined by:

$$\textit{Minimize} \sum_{k \in [\![0, V-1]\!]} \Delta_{v_k} \quad [5.15]$$

Under the constraints:

$$\forall i \in [\![1, M]\!], \quad e_{m0i} = \overrightarrow{m_0}(i) \quad [5.16]$$

$$\forall i \in [\![1, M]\!], \quad e_{mVi} = \overrightarrow{m_f}(i) \quad [5.17]$$

$$\forall j \in [\![1, N]\!], \quad e_{r0j} = \overrightarrow{r_0}(j) \quad [5.18]$$

$$\forall j \in [\![1, N]\!], \quad e_{rVj} = \overrightarrow{r_f}(j) \quad [5.19]$$

$$\forall k \in [\![0, V-1]\!], \ \forall j \in [\![1, N]\!], \quad B.\alpha_{kj} - e_{rkj} \leq B - 1 \quad [5.20]$$

Figure 5.9. *Integer programming model*

$\forall k \in [\![0, \mathrm{V}-1]\!]$,	$\forall j \in [\![1, \mathrm{N}]\!]$,	$e_{rkj} - B.\alpha_{kj}$	$\leq$	0	[5.21]
$\forall k \in [\![0, \mathrm{V}-1]\!]$,	$\forall j \in [\![1, \mathrm{N}]\!]$,	$B.\alpha_{kj} - e_{rkj} + \Delta_{\mathrm{v_k}}$	$\leq$	$B-1$	[5.22]
$\forall k \in [\![0, \mathrm{V}-1]\!]$,	$\forall j \in [\![1, \mathrm{N}]\!]$,	$e_{rkj} - \Delta_{\mathrm{v_k}} - B.\alpha_{kj}$	$\leq$	0	[5.23]
$\forall k \in [\![0, \mathrm{V}-1]\!]$,	$\forall j \in [\![1, \mathrm{N}]\!]$,	$e_{rkj} - \Delta_{\mathrm{v_k}} - \beta_{kj}$	$\leq$	0	[5.24]
$\forall k \in [\![0, \mathrm{V}-1]\!]$,	$\forall j \in [\![1, \mathrm{N}]\!]$,	$\beta_{kj} - B.\alpha_{kj}$	$\leq$	0	[5.25]
$\forall k \in [\![1, \mathrm{V}]\!]$,	$\forall j \in [\![1, \mathrm{N}]\!]$,	$\beta_{kj} + B.\alpha_{kj} - e_{rkj} + \Delta_{\mathrm{v_k}}$	$\leq$	B	[5.26]
$\forall k \in [\![1, \mathrm{V}]\!]$,	$\forall j \in [\![1, \mathrm{N}]\!]$,	$e_{mki} - e_{m(k-1)i} + \sum_{c=1}^{\mathrm{N}} C_{ic}^{-} \cdot \sigma_{kc} - \sum_{c=1}^{\mathrm{N}} C_{ic}^{+}(a_{(k-1)c} - \alpha_{(k-1)c})$	$=$	0	[5.27]
$\forall k \in [\![1, \mathrm{V}]\!]$,	$\forall j \in [\![1, \mathrm{N}]\!]$,	$e_{rkj} - \beta_{(k-1)j} - \vec{d}(j).\sigma_{kj}$	$=$	0	[5.28]
$\forall k \in [\![1, \mathrm{V}]\!]$,	$\forall j \in [\![1, \mathrm{N}]\!]$,	$\sigma_{kj} + \alpha_{(k-1)j}$	$\leq$	1	[5.29]
$\forall k \in [\![1, \mathrm{V}]\!]$,	$\forall j \in [\![1, \mathrm{N}]\!]$,	σ_{kj}	$\in$	$\{0,1\}$	[5.30]
$\forall k \in [\![0, \mathrm{V}-1]\!]$,	$\forall j \in [\![1, \mathrm{N}]\!]$,	a_{kj}	$\in$	$\{0,1\}$	[5.31]
$\forall k \in [\![0, \mathrm{V}-1]\!]$,	$\forall j \in [\![1, \mathrm{N}]\!]$,	α_{kj}	$\in$	$\{0,1\}$	[5.32]
$\forall k \in [\![0, \mathrm{V}-1]\!]$,	$\forall j \in [\![1, \mathrm{N}]\!]$,	β_{kj}	$\in$	$\mathbb{N}$	[5.33]
$\forall k \in [\![0, \mathrm{V}]\!]$,	$\forall j \in [\![1, \mathrm{N}]\!]$,	e_{rkj}	$\in$	$\mathbb{N}$	[5.34]
$\forall k \in [\![0, \mathrm{V}]\!]$,	$\forall i \in [\![1, \mathrm{M}]\!]$,	e_{mki}	$\in$	$\mathbb{N}$	[5.35]
$\forall k \in [\![0, \mathrm{V}-1]\!]$,		$\Delta_{\mathrm{v_k}}$	$\in$	$\mathbb{N}$	[5.36]

Figure 5.9. *(Continued)*

5.4.3. *Numerical experiments*

Experiments have been done on a dual-processor Intel Xeon 64-bit with 3 GHz and 4 GB RAM, under Ubuntu Linux, using the integer linear programming library CPLEX 5.1. Each time, we look for the fastest executions leading from a given initial marking to the final marking. The computation time is always indicated in seconds. It corresponds to the duration of model construction, the duration of some intermediate informative displays and the duration of the resolution itself.

When we do not know the number of steps needed to solve each problem beforehand, we perform a series of resolutions on the model in Figure 5.9, gradually increasing the size of the search horizon (i.e. the number K of steps) from $k = 1$, until the *minimum reachability horizon* wherein a result exists, denoted by K_{min}. On this step, we stop the search as soon as a feasible solution is obtained, without trying to reach the optimum.

For each resolution, to reduce the search space, we introduce a "cut" prohibiting the firing of empty steps. However, we authorize the last step to be empty in order to be able to reach the desired final state for which the residual duration vector is usually zero. Formally, this cut is expressed by equation [5.37]. The experimental results show that it greatly reduces resolution time.

$$\forall k \in [\![1, V-,]\!], \sum_{j=1}^{N} \sigma_{kj} > 0 \qquad [5.37]$$

Another cut will also be tested, characterizing the firing of each transition a minimum number of times, given by the value $\vec{\varphi} \in \mathbb{N}^N$. This cut is expressed by inequality [5.38a]. Such a vector $\vec{\varphi}$ can frequently be obtained using a preliminary analysis based on the resolution of the equation of state or the untimed reachability problem. It can also be the result of empirical considerations on the physical model.

$$\forall j \in [\![1, \mathbb{N}]\!], \sum_{k=1}^{V} \sigma_{kj} \geq \varphi_j \qquad [5.38a]$$

$$\forall j \in [\![1, \mathbb{N}]\!], \sum_{k=1}^{V} \sigma_{kj} = \varphi_j \qquad [5.38b]$$

Using an inequality may not necessarily constrain the number of firings of all transitions, since the inequality is trivially true for the index j when $\varphi j = 0$. When the number of firings for each transition is well known and all that remains is to schedule these transitions by assigning firing dates, we can replace inequality [5.38a] with the alternative equation [5.38b]. This will further improve the response time of solvers, as discussed below.

Of course, this inequality is not necessary to ensure accuracy of the model, it simply allows us to highlight the possibility of searching for properties by adding a given set of constraints onto firings or intermediate markings characterizing these properties on the model. These constraints can be equalities or inequalities, can focus on steps $\overrightarrow{\sigma_k}_{,k\in[\![1,\mathrm{V}]\!]}$, markings $\overrightarrow{em_k}_{,k\in[\![0,\mathrm{V}]\!]}$ or even intermediate delays $\Delta_{vk,k\in[\![0,\mathrm{V}-1]\!]}$, but they must be linear so as not to cause efficiency loss in solving the linear programming problem.

5.4.4. *Study of the illustrative example of Figure 5.5*

In Table 5.6, we present the results of the study of the reachability problem for the example in Figure 5.5, with an initial state of the form $(J \cdot p_1,\ J \cdot p_4,\ r_1,\ r_2)$, on which we vary the quantity J of tokens in places p_1 and p_4. The goal is to make tokens of places p_1 and p_4 move as quickly as possible to places p_3 and p_6, the final state remaining identical to the initial state otherwise.

In column T_1, we indicate the duration of the incremental search for $\mathrm{K}_{\min}$. We also show the characteristics of the problem for a number of steps equal to the value of $\mathrm{K}_{\min}$ found: number of variables V and constraints C of the solved model, duration T_2 of the resolution, value O of the objective function. The dates T_3 and T_4 respectively correspond to the use of cuts defined by the equations [5.37] and [5.38b], through which we force each transition to be fired J times. This first example gives us the opportunity to compare the influence of the values of constant B (100 to 10,000) on the speed of resolution.

As can be seen, the computation time is very important for a small number of variables and constraints which confirms the difficulty of the problem.

The influence of the value of B is difficult to define, with no value dominating any other in a sensitive manner. However, for more complex problems, low values give better results. In the following, we thus use a low value of B while maintaining compatibilty with the model, for example, $B = 100$.

As previously stated, the use of cuts [5.37] and [5.38b] helped to significantly improve computation times for low values of the parameter B.

J		1	2	3	4	5
K_{min}		3	5	7	9	11
V		99	157	215	273	331
C		156	244	332	420	508
O		9 t.u.	18 t.u.	27 t.u.	36 t.u.	45 t.u.
T_1		0.01	0.04	0.93	11.04	738.43
T_2	$B =$	0.01	0.02	1.74	30.60	7521.43
T_3	100	0.00	0.05	1.49	57.59	3327.89
T_4		0.00	0.03	0.85	33.97	560.72
T_1		0.01	0.05	0.38	5.46	1232.81
T_2	$B =$	0.00	0.02	0.56	35.75	N.C.
T_3	1 000	0.00	0.01	0.22	2752	10441.06
T_4		0.00	0.03	0.32	64.92	484.62
T_1		0.00	0.05	1.47	25.41	48624.26
T_2	$B =$	0.01	0.03	1.48	56.51	N.C.
T_3	10 000	0.01	0.02	0.26	41.66	10215.44
T_4		0.01	0.03	0.42	8.06	18299.37

Table 5.6. *Numerical results for the example in Figure 5.5*

5.4.5. *Conclusion and future work*

The explored cutting mechanism achieves significant improvements in reducing the solution search space, even when we know the number of firings of each transition.

Some experiments were led on more complex problems like the railway system presented above. Results show that our approach must be improved to solve non-trivial problems on instances of academic size. For instance, we show in Table 5.7 the characteristics of a resolution corresponding to the Petri net in Figure 5.7, with one train moving in each direction. Times T_1 and T_2 respectively denote the time needed to find the minimum search depth K_{min}, and the resolution time when the search depth is fixed to this value from the beginning.

The computational times obtained are quite good and validate the proposed modeling, but they become prohibitive (we stopped our search after more than two hours) when the number of trains in the Petri net grows. Several approaches are currently being led to improve this work, following various directions: development of cutting techniques for mathematical programming, use of constraint programming approaches or event-dedicated heuristics.

In the next section, we present an alternative modeling approach based on a subclass of Timed Petri nets to address the same kind of problems.

K_{min}	V	C	O	T_1	T_2
21	2735	4461	60	570.81	121.62

Table 5.7. *Scheduling with time windows for the example of Figure 5.7*

5.5. A (max,+) approach to scheduling

The linear nature of a simple events graphs has led to the development of algebraic structures that are dioids, (min,+) or (max,+). These structures are in default when event graphs are weighted, and even more so when the Petri net contains indeterminisms related to resource sharing. In this chapter, only Petri nets whose indeterminisms are related to the sharing of resources are studied. We demonstrate that it is possible to transpose a study of this type of Petri net on the events graph and a system of associated equations.

5.5.1. *Introduction and production hypotheses*

Here we address the problem of indeterminism in Petri nets, especially those relating to resource sharing. For certain configurations, flexible workshops contain multi-tasking machines shared between different processes to be completed. It is therefore necessary to order them in sequence to determine piloting of the workshop.

In previous work, this problem was solved using investment policies or routing functions [OHL 94]. Our goal here is to produce every possible sequence of resources to ensure an optimal final sequence. Our approach involves decomposing the resource scheduling problem in two steps:

– Firstly, the study of logical behavior that aims to abstract all possible sequences in a system of equations and inequalities whose variables are the initial markings of places.

– Secondly, an assessment of the cycle time of the obtained solutions can help obtain an optimal solution for a fixed search horizon.

In this chapter, we propose a new approach under the following operating assumptions:

– The number of operations that a piece must undergo is known and deterministic, only machinery shared between several tasks induce structural conflicts.

– The number of assignments for each machine cycle was determined in an earlier optimization flow phase [KOR 98; CAM 97; OHL 95]. The resulting earliest optimization phase model is a Petri net of a particular type as it is relatively close to the model of the weighted events graph. The weighting most often arises from the consideration of lots of components or parts, and also complex assembly processes. Other than this weighting, the only other residual nonlinearity results from conflicts of access to resources whose allocation is not determined at this level of study.

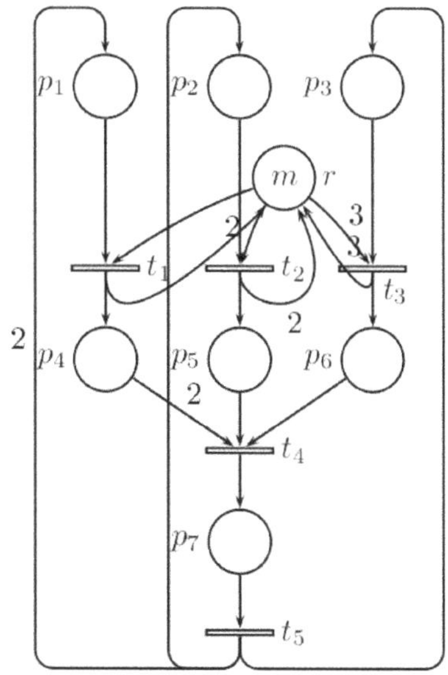

Figure 5.10. *Illustrative example*

As an example, consider the case of the production range described in Figure 5.10 The production chain allows the manufacturing of a product produced by an assembly of three types of components: two of type A_1, one of each type A_2 and A_3.

The first component A_1 is produced by processing the part located at p_1 using a copy of resource r, which is symbolized by firing transition t_1. Manufacturing the second component is performed on transition t_2 using two copies of resource r. The last component is processed with three type r resources and is symbolized by firing transition t_3. An assembly of two A_1 parts, an A_2 part and an A_3 part is then performed at transition t_4. Finally, the result of the assembly is controlled, symbolized by the firing of transition t_5 and obtainment of the final product. The model is virtually fed back to take the system's cyclic production into account.

This example illustrates the practical interest in weighting of arcs, as much to characterize the number of components to be assembled as to specify the number of resources allocated to a particular activity.

We will address the transformation problem of sharing multiple resources among different tasks where arcs are weighted and the firing of transitions is not necessarily unitary. This problem has already been addressed under more restrictive assumptions; sharing a resource without weighting and with unitary firings of transitions in conflict is discussed in [BEN 99], sharing several identical resources is addressed in [SIL 02], and sharing resources with multiple firing of transitions in conflict in [TRO 02b]. In this chapter, we consider the case of adding a weight on arcs associated with resource sharing.

The advantage of our approach lies in the possible extension of the field of application of formalisms (max,+) and (min,+), developed for performance evaluation of simple event graphs [COH 89; LAH 99; GAU 95]. In this regard, the exposed transformation will allow obtainment of a simple, or possibly weighted, graph of events. In the latter case, a linearization technique described in [JUA 01; TRO 02a] will use the dioid (min,+) to study the behavior of the logic model by propagating an abstract initial marking.

The developed method will therefore address the problem of optimal scheduling of resources under the assumption of determined numbers of transitions' firings with access conflict. The aim is to study optimal scheduling in two steps:

– a study of logical behavior: to obtain possible sequencing of resources;

– a study of temporal behavior: to evaluate the cycle time associated with each schedule.

Thus the purpose of this chapter is to transform the resource scheduling problem into an initial marking search problem in a weighted events graph. Assuming earliest firing, a study of logical behavior can then be performed on the event graph.

In this sense, we propose a three-part plan. As a first step, we propose the construction of a weighted event graph obtained from an initial Petri net. The goal is to make the network deterministic, a property required for an algebraic behavior approach. In a second step, the introduction of a system of equations and inequalities justifies the equivalence between all possible resource sequences and all the behaviors generated by the resulting system solutions. The third part is devoted to an application of this transformation method.

5.5.2. *Construction of a simple event graph associated with the initial model*

5.5.2.1. *Construction method*

We first propose to construct an events graph derived from the initial model that we describe, using a specific initial marker and any potential transition sequences that are fireable when the transitions share m identical resources. The purpose here is to keep all potential m resource sequence opportunities in the construction of a simple events graph. In this sense, we propose to impose a partial order on transitions, thus introducing two places for each pair of transitions; marking these places will then define the firing order between two transitions t_i and t_j. To this end, we propose the following transformation (Figure 5.11).

DEFINITION 5.1.– We use the abbreviation TGES for the Petri net obtained by processing each place r that represents a shared resource, this transformation is described below:

– We eliminate the initial position r associated with sharing of resources.

– Each transition t* is duplicated as many times as necessary depending on the value of the number of firings associated with transition t in a cycle. (We obtain a set of observed transitions, denoted T').

– $\forall i,j \in [\![0,|T'|]\!]$, $1 \leq i < j \leq n$, we introduce two places represented by p_{ij} and p_{ji} between the two transitions t_i^* and t_j^*, whose marking is limited to one.

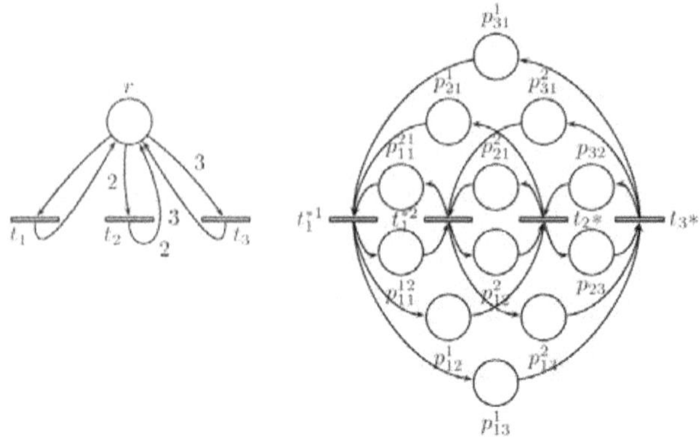

Figure 5.11. *Resource sharing*

The above construction can transform the initial Petri net into a possibly weighted events graph, depending on the nature of the initial Petri net.

Using this transformation, we get a simple or weighted events graph depending on whether the nature of the initial Petri net is simple or weighted. The first benefit of such a transformation is the resulting graph's determinism; the second benefit is the linear behavior (min,+) of such a system provided that the initial Petri net is simple. In the case of a weighted events graph, we previously demonstrated [JUA 01; TRO 02a] how it is possible to linearize the obtained model to finally get a graph.

At this stage of the study, the previous construction can transform a Petri net into a potentially weighted events graph. The purpose of the following paragraph is to establish the bi-univocal correspondence between the set of possible sequences from many copies of the resource and all behaviors induced by the equations system and inequality solutions.

5.5.2.2. *Assessment of the number of places created*

The previously proposed construction transforms resource sharing into an unweighted events graph. The transformation induces the addition of a number of places proportional to the number of conflicting transitions in the initial Petri net. Therefore, the number of places added during the construction can be evaluated.

Let I be the set of indices of conflicting transitions; λ_i the number of firings of transition t_i during a production cycle; np the number of places created, is evaluated according to the following formula:

$$n_p = 2 \times \left(\sum_{i \in I} \lambda_i\right)!$$

where the place containing the resources was removed during construction. We conclude that the number of places in the Petri net is increased by n_p - one place. Addition to places and related arcs generates different circuits in the Petri net. To assess the number of circuits, we propose to calculate the number of TGES elementary circuits previously obtained. First though, we must note that the graph is strongly connected.

By definition, a path joining any two TGES transitions exists. In addition, the particular TGES construction confers the property of only having one arc joining two transitions. We deduce that the number of elementary paths containing n TGES transitions is equal to the cardinality of the set of permutations of n modulo elements of circular permutations of these transitions.

A basic TGES circuit is characterized by the transitions within it. The circuits can then be represented by transition p-uplets. However, several p-uplets may represent the same circuit. Indeed, the p-uplets representing the same circuit are obtained by circular permutation of the elements.

We deduce the number of elementary circuits present in TGES:

$$n_c = \sum_{i=2}^{|T'|} \frac{\mathrm{A}_{|T'|}^{i}}{i}$$

Where $\mathrm{A}_{|T'|}^{i}$ represents the number of arrangements of i elements from $|T'|$. As proposed for the 20 basic circuits' example.

5.5.3. *Resolution of resource sharing*

We focus this study on the transformation of a structural conflict signal generated by resource sharing. In the previous section, we saw that transformation of a structural conflict can lead to an events graph. The purpose of this section is to abstract all possible sequencing of resources. Each sequence can be characterized by an initial TGES marking obtained by the construction method described above. This paragraph is devoted to the search for a system of inequalities whose set of solutions correspond to those obtained from the search for a set of initial markings. Each initial marking then generates any possible sequence of resources and vice versa.

This section is divided into three parts. In the first part, we present a method for obtaining a system of inequalities where an initial marking solution corresponds to a given sequence of resources. In the second part, we extend the system of inequalities obtained above to ensure the vitality of an initial marking. To conclude, we prove that all possible sequences of resources can be obtained using an initial marking of the global system of TGES inequality solutions thus generated.

5.5.3.1. *Obtaining a system of inequalities*

Due to the transformation referred to in paragraph two, we are studying the case of sharing m identical copies of a resource r between d tasks, represented by a set of transitions $T' = \{t_1^*, ..., t_d^*\}$, crossed only once per cycle. The purpose of this section is to produce a set of relations that must verify an initial marking to generate any of the resource sequences. More generally, we seek to obtain a system of inequalities such that each initial marking corresponds to a sequence of resources, a system solution. Assuming parallel execution, we determine in fact a sequence of achievable steps [BEN 00] in terms of specified assumptions about how the TGES functions. In this regard, we give the definition of a sequence in the context presented above:

DEFINITION 5.2.– A sequence, in a network containing multiple resources shared between d tasks, is a sequence of steps S_1, S_2,..., S_k. Each transition occurs once and only once in the sequence and the maximum step size cannot exceed m, the number of identical machines[1].

After introducing the necessary formalism, we produce a method for obtaining an initial marking generating any of the specified timings. The initial sought marking must integrate the fact that m resources are shared. As a result, several transitions can be crossed simultaneously: the amount depends on the number of resources necessary to perform this step. Therefore, we define sets of transitions based on the number of resources used in case of parallel firing of all transitions.

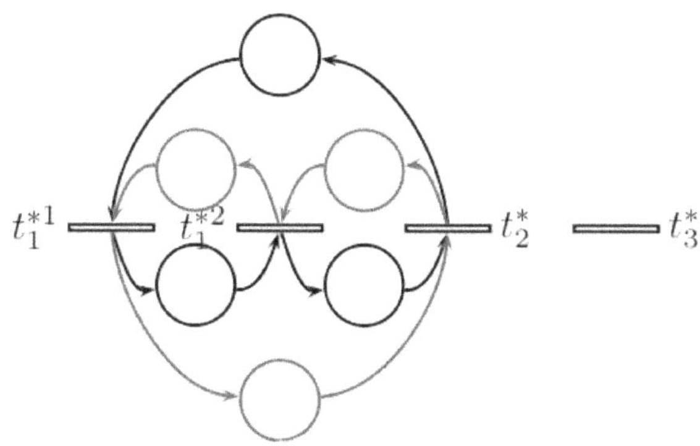

Figure 5.12. *Example of paths*

DEFINITION 5.3.– We denote T_i^* as the subset $\{t_{i_1}^* \ldots t_{i_{\alpha+1}}^*\}$ of transitions of T' such that:

$$(\forall j) : 1 \leq j \leq \alpha + 1, \sum_{k=1;k\neq j}^{\alpha+1} W(r, t_{i_k}) \leq m \text{ and } \sum_{k=1}^{\alpha+1} W(r, t_{i_k}) > m \quad [5.39]$$

1 In the case of sharing a single resource, the sequence of steps becomes a sequence of transitions.

where $W(r, t_{i_k})$ represents the weight of the arc joining place r to transition t_{ik}.

where $W(r, t_{i_k})$ represents the weight of the arc joining place r to transition t_{ik}.

The above definition allows us to consider subsets of transitions T_i^* of all TGES transitions. Inequalities [5.39] affirm that α transitions of T_i^* can be crossed but simultaneous firing of $\alpha + 1$ transitions of T_i^* is impossible. The idea here is to prohibit the simultaneous firing of all transitions contained in a set of T_i^*, which we state with the following lemma:

LEMMA 5.1.– Let M be an initial TGES marking, if it satisfies the following conditions:

$$\forall\, T_i^* = \{t_{i_1}^*, \cdots, t_{i_{\alpha+1}}^*\}, \left(\sum_{j=1}^{\alpha} m_{p_{i_{j+1}i_j}}\right) + m_{p_{i_1 i_{\alpha+1}}} \leq \alpha \qquad [5.40]$$

then simultaneous firing of all transitions of T_i^* is impossible in a single step.

PROOF.– Let a subset of type T_i^* transitions $= \{t_{i_1}^*, \cdots, t_{i_{\alpha+1}}^*\}$, these transitions satisfy inequalities [5.39] by hypotheses. So the simultaneous firing of all transitions T_i^* cannot occur.

In this regard, we carry out type T_i^* sets, then form the set of elementary circuits based on transitions contained in the T_i^* sets. Let us consider a circuit of this type, for which all contained transitions firing in parallel require a greater number of resources than the present resources (inequality [5.39]). Therefore, it is not appropriate to simultaneously validate all T_i^* transitions, or α + 1 transitions. However, any strict subset of transitions can be validated during a step (inequality [5.39]), or during α transitions. Therefore, at least one place in this circuit should not be marked, which implies that all transitions cannot be crossed in the first step. Consequently, it suffices to place at most α tokens among $\alpha + 1$ places (whether the marking is 0 or 1) of the considered elementary circuit.

Inequality [5.40] holds true throughout the network evolution. Places contained in the previously described circuits form invariants of places. In fact, every transition has an upstream and downstream place. Therefore, the firing of transitions does not alter the amount of place markings. Therefore, markings check inequality [5.40] during the operation of the Petri net, so all considered transitions can never be crossed. For example, consider for TGES shown in Figure 5.10: a subset of transitions, answering to the criteria in definition 5.3, is given by $T_1^* = \{t_1^{*1}, t_1^{*2}, t_2^*\}$. We obtain two basic paths consisting of three transitions from the set T_1^* represented in Figure 5.12. From lemma 5.1, we can deduce that for each circuit, one of the places should not be marked to avoid parallel firing of transitions t_1^{*1}, t_1^{*2} and t_2^*.

Lemma 5.1 expresses a condition sufficient for an initial TGES marking to induce a behavior whereby steps consuming an amount of resources smaller than or equal to m, the number of available resources. This result leads to a system of inequalities such as those in [5.40]. This system of inequations contradicts the TGES marking of places in order to obtain an initial marking generating a resource sequence while validating the specifications associated with transition firing. The system of inequalities obtained from lemma 5.1 affirms that a marking solution generates a correct behavior in terms of resource use. Consequently, the set of initial markings generating correct behavior is included in the set of initial solution markings of the system of inequalities. The purpose of this section is to reduce the set of solutions for obtaining the reciprocal inclusion.

5.5.3.2. *Integration of liveliness conditions*

Our goal now is to associate an additional system of inequalities ensuring the liveliness of the model to the system obtained using the generic equation [5.40]. Theorem 5.1 allows us to assert that the initial marking solution of the system of inequalities generates a correct behavior or a blockage. The purpose of this section is to eliminate the solutions causing a blockage of the subnet.

Our reasoning is based on the following theorem:

THEOREM 5.1.– If M is an initial TGES marking checking relations [5.40] then either marking induces a blockage, or there is a cyclic sequence σ such that $M[\sigma > M$.

PROOF.– Let M be a marking satisfying relation [5.40]. We assume that the marking M does not induce blocking, that is to say we assume the existence of a live transition, which we denote t_i. We further assume the existence of a sequence of steps such that:

$$S_1S_2 \ldots S_k \text{ such that } M\,[S_1S_2 \ldots S_k\rangle$$

Firing transition t_i of this circuit involves the consumption of tokens upstream of transition places. But:

$$\forall j \neq i, \{p_{ji}\} = {}^{\bullet}t_i \cap {}^{\bullet}t_j \qquad [5.41]$$

Firing t_i only becomes possible when the upstream places are marked again. Equation [5.41] holds that firing of t_i is only possible after all TGES transitions have been crossed. Thus, index k may be chosen such that the sequence of steps $S = S_1S_2 \ldots S_k$ is the maximum possible, and contains one and only one occurrence of each transition. Transition t_i is assumed to be live and can only be crossed if the others have been fired. So there are two possibilities: either S contains all the transitions and the proof is complete, or S does not contain all the transitions and we deduce the existence of a non-fireable transition, which implies non-crossability of transition t_i of the marking obtained after firing sequence S. But given the maximality of sequence S and the liveliness of transition t_i, we conclude that this is impossible. As a result, S is a sequence of steps with all transitions and theorem 5.1 requires that the initial marking satisfies an inequality which prohibits the simultaneous firing of more than α transitions. In conclusion, the sequence of steps $S = S_1S_2 \ldots S_k$ is a cyclic sequence.

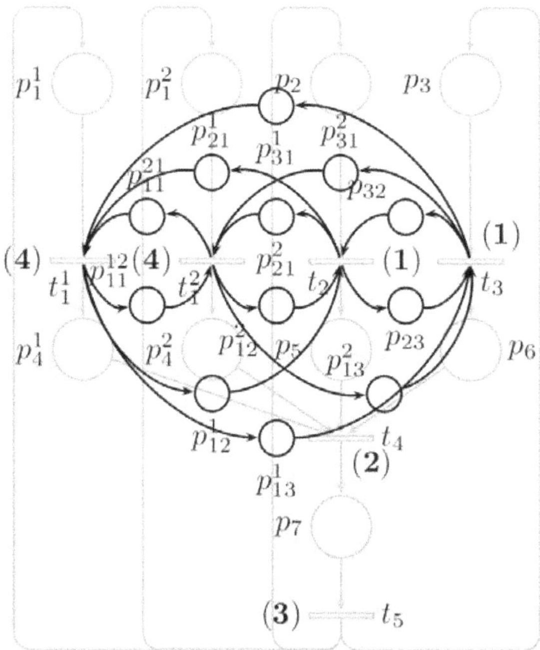

Figure 5.13. *The resulting events graph*

The main idea in this section is to filter solutions that correspond to a TGES blockage. The guarantee for the liveliness of solutions is therefore to avoid blockages so that only the possible resource sequences remain. With reference to previous work [COM 71], we recall the necessary and sufficient conditions for liveliness as follows:

THEOREM 5.2.– For a simple events graph to be live, at least one token in each elementary circuit of the network must exist.

It is then necessary to describe the TGES elementary circuits to ensure the liveliness of a marking using theorem 5.2. That is to say, the sum of markings of places contained in each elementary circuit must be greater than or equal to one. In addition, a basic elementary circuit may not contain more transitions than the total number of TGES transitions, under penalty of repetition. So the inequalities cannot contain more unknowns than elements in the set *T'*; in fact, an inequality of the above type contains markings of intermediate places between two transitions of a path containing a subset of T' without repetition. Constraints allow enunciation of the following theorem:

COROLLARY 5.1.– The system of inequalities (definition 5.3), increased by the liveliness inducing system, only captures all possible cyclic sequences.

Consider a weighted Petri net for which transition firing constraints have been expressed and whose indeterminisms only come from shared resources. Corollary 5.1 allows us to assert that it is possible to transform it into a weighted events graph while preserving all possible behaviors of the Petri net.

5.5.4. *Application*

Let us apply the above theory to the example in Figure 5.13 from the beginning of this chapter. We consider the case of three resources (m = 3). In order to assess the temporal behavior of the network, delays associated with tokens are added to the previous model (Figure 5.10).

5.5.4.1. *Study of logical behavior*

The first step involves duplicating transition t_1 into two transitions t_1^{*1} and t_2^{*2}, whose frequency is of one firing per cycle. The Petri net is transformed by removing sharing of resources and constructing the TGES, shown in black in Figure 5.14. We then resolve resource sharing using the method described in section 5.5.3.

We need to add the systems of inequalities related to resources' sharing and liveliness conditions into the TGES. The first step is to determine all subsets of $T' = \{t_1^{*1}, t_1^{*2}, t_2^{*}, t_3^{*}\}$ that respect the conditions in definition 5.3. We obtain:

$$\begin{cases} \{t_2^*, t_3^*\} \\ \{t_1^{*1}, t_3^*\} \\ \{t_1^{*2}, t_3^*\} \\ \{t_1^{*1}, t_1^{*2}, t_2^*\} \end{cases} \qquad \begin{cases} m_{p_{32}} + m_{p_{23}} \leq 1 \\ m_{p_{31}^{1}} + m_{p_{13}^{1}} \leq 1 \\ m_{p_{31}^{2}} + m_{p_{13}^{2}} \leq 1 \end{cases} \qquad \begin{cases} m_{p_{21}^{21}} + m_{p_{12}^{1}} + m_{p_{21}^{2}} \leq 2 \\ m_{p_{11}^{12}} + m_{p_{12}^{2}} + m_{p_{21}^{1}} \leq 2 \end{cases} \qquad [5.42]$$

Theorem 1 allows us, from subsets of T' given in the system [5.42], to produce a system that ensures the validity of the obtained solutions, meaning that the initial markings obtained generate a sequence of resources. This results in the following system of inequalities ensuring the liveliness of sequencing:

$$\begin{cases} m_{p_{11}^{21}} + m_{p_{11}^{12}} \geq 1 \\ m_{p_{21}^{2}} + m_{p_{12}^{2}} \geq 1 \\ m_{p_{32}} + m_{p_{23}} \geq 1 \end{cases} \qquad \begin{cases} m_{p_{21}^{1}} + m_{p_{12}^{1}} \geq 1 \\ m_{p_{31}^{2}} + m_{p_{13}^{2}} \geq 1 \\ m_{p_{31}^{1}} + m_{p_{13}^{1}} \geq 1 \end{cases}$$

$$\begin{cases} m_{p_{11}^{21}} + m_{p_{12}^{1}} + m_{p_{21}^{2}} \geq 1 \\ m_{p_{11}^{12}} + m_{p_{12}^{2}} + m_{p_{21}^{1}} \geq 1 \\ m_{p_{11}^{21}} + m_{p_{13}^{1}} + m_{p_{31}^{2}} \geq 1 \\ m_{p_{11}^{12}} + m_{p_{13}^{2}} + m_{p_{31}^{1}} \geq 1 \\ m_{p_{21}^{1}} + m_{p_{13}^{1}} + m_{p_{32}} \geq 1 \\ m_{p_{12}^{1}} + m_{p_{23}} + m_{p_{31}^{1}} \geq 1 \\ m_{p_{21}^{2}} + m_{p_{13}^{2}} + m_{p_{32}} \geq 1 \\ m_{p_{12}^{2}} + m_{p_{23}} + m_{p_{31}^{2}} \geq 1 \end{cases} \qquad \begin{cases} m_{p_{11}^{21}} + m_{p_{21}^{2}} + m_{p_{32}} + m_{p_{13}^{1}} \geq 1 \\ m_{p_{11}^{12}} + m_{p_{12}^{2}} + m_{p_{23}} + m_{p_{31}^{1}} \geq 1 \\ m_{p_{12}^{1}} + m_{p_{21}^{2}} + m_{p_{13}^{2}} + m_{p_{31}^{1}} \geq 1 \\ m_{p_{21}^{1}} + m_{p_{12}^{2}} + m_{p_{31}^{2}} + m_{p_{31}^{1}} \geq 1 \\ m_{p_{13}^{1}} + m_{p_{31}^{2}} + m_{p_{12}^{2}} + m_{p_{21}^{1}} \geq 1 \\ m_{p_{31}^{1}} + m_{p_{13}^{2}} + m_{p_{21}^{2}} + m_{p_{12}^{1}} \geq 1 \end{cases}$$

[5.43]

This transformation leads to the production of the weighted events graph shown in Figure 5.14. The logical behavior of the Petri net can be studied using the dioid (min,+) [TRO 02a]. We then obtain the following linear system of equations (44) (min,+):

$$\begin{cases} n_1^1(k+1) = min\left(n_1^2(k) + m_{p_{11}^{21}}, n_2\,(k) + m_{p_{21}^{1}}, n_3\,(k) + m_{p_{31}^{1}}, n_5\,(k) + m_{p_1^1}\right) \\ n_1^2(k+1) = min\left(n_1^1(k) + m_{p_{11}^{12}}, n_2\,(k) + m_{p_{21}^{2}}, n_3\,(k) + m_{p_{31}^{2}}, n_5\,(k) + m_{p_1^2}\right) \\ n_2\,(k+1) = min\left(n_1^1(k) + m_{p_{12}^{1}}, n_1^2(k) + m_{p_{12}^{2}}, n_3\,(k) + m_{p_{32}}, n_5\,(k) + m_{p_2}\right) \\ n_3\,(k+1) = min\left(n_1^1(k) + m_{p_{13}^{1}}, n_1^2(k) + m_{p_{13}^{2}}, n_2\,(k) + m_{p_{23}}, n_5\,(k) + m_{p_3}\right) \\ n_4\,(k+1) = min\left(n_1^1(k) + m_{p_4^1}, n_1^2(k) + m_{p_4^2}, n_2\,(k) + m_{p_5}, n_3\,(k) + m_{p_6}\right) \\ n_5\,(k+1) = n_4\,(k) + m_{p_7} \end{cases}$$

[5.44]

Where the $n_i(k)$ variable characterizes the state of the counter of occurrences of start of transition i for step k. We obtain the following recurrent matrix equation deducted from the previous system:

$$N(k+1) = B(M(0)) \otimes N(k)$$

Where the operator $\otimes$ denotes the matrix product in the sense of the dioid (min, +) and matrix $B(M(0))$, derived from the system, is a matrix set by the initial marking described below:

$$\begin{pmatrix} +\infty & m_{p_{11}^{21}} & m_{p_{21}^{1}} & m_{p_{31}^{1}} & +\infty & m_{p_{1}^{1}} \\ m_{p_{11}^{12}} & +\infty & m_{p_{21}^{2}} & m_{p_{31}^{2}} & +\infty & m_{p_{1}^{2}} \\ m_{p_{12}^{1}} & m_{p_{12}^{2}} & +\infty & m_{p_{32}} & +\infty & m_{p_{2}} \\ m_{p_{13}^{1}} & m_{p_{13}^{2}} & m_{p_{23}} & +\infty & +\infty & m_{p_{3}} \\ m_{p_{4}^{1}} & m_{p_{4}^{2}} & m_{p_{5}} & m_{p_{6}} & +\infty & +\infty \\ +\infty & +\infty & +\infty & +\infty & m_{p_{7}} & +\infty \end{pmatrix}$$

Therefore, the study of logical behavior of the start of the Petri net boils down to the study of a system of equations whose solutions characterize the initial marking of a weighted events graph (system [5.44]), and of a system of equations and inequalities characterizing the correct sequencing of the available copies of resource r (system [5.42] and [5.43]). Finally, to obtain optimal scheduling on a timed model, it is necessary to estimate the cycle time associated with the obtained sequences.

5.5.4.2. *Resolution and evaluation of performances*

In this section, we propose a solution for previously established equations and the systems of inequalities by searching for a proper schedule of Petri nets obtained by cycle time evaluation. This resolution is made in three steps:

– set a search horizon: the horizon is reflected by the number of steps needed to achieve a production cycle;

– solve the obtained final system whose solutions are the initial markings of the transformed Petri net;

– evaluate the cycle time associated with the solutions obtained.

In the first step, the search horizon depends on the number of resources assigned to the operations to be performed. For our example shown in Figure 5.14, type *r* resource is shared between three tasks. With the production assumptions mentioned above, an admissible schedule of three copies of the resource induces an operation in a minimum of *ns* steps, evaluated as follows:

$$n_s = \frac{\sum_{t\in T}(\text{number of firings of transition } t \text{ per cycle}) \times W(r,t)}{\text{number of resources } (m)}$$

It is therefore also necessary to consider at least three steps for solving the system of equations and inequalities.

Let us conduct a search for a solution for horizon *ns*, set to three steps. We must solve the matrix equation:

$$N(k+3) = B(M(0)) \otimes N(k)$$

This corresponds to a cyclic system in three steps. The resolution is made using the CPLEX software. The system of equations and inequalities are implemented, which allows us to obtain solutions through enumeration. For example:

$$m_{p_1^1} = m_{p_1^2} = m_{p_6} = m_{p_7} = m_{p_{11}^{21}} = m_{p_{21}^1} =$$

$$m_{p_{31}^1} = m_{p_{21}^2} = m_{p_{11}^{12}} = m_{p_{31}^2} = m_{p_{32}} = 1$$

Other non-listed places are not marked. The initial marking generates a cyclic behavior in the three steps given below:

Step 1: $t_1^1//t_1^2//t_5$
Step 2: t_2
[5.45]
Step 3: t_3 // t_4

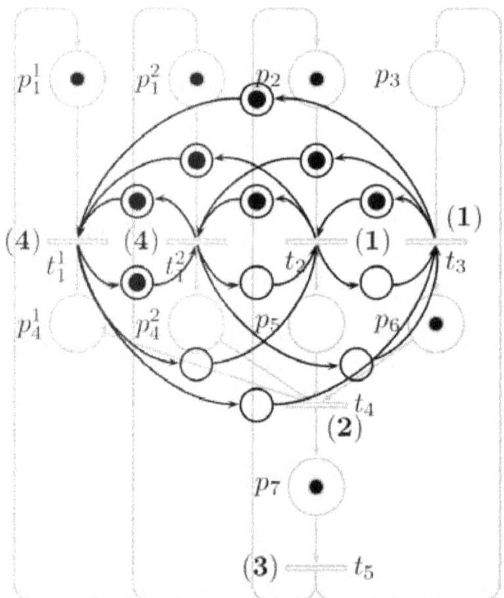

Figure 5.14. *The resulting weighted events graph*

The times illustrated in Figure 5.14 allow an assessment of the cycle time associated with the previously proposed sequence. We obtain a cycle time of 7 time units.

Our search was carried out over a horizon of three steps, the goal was to promote parallelism in the tasks to minimize the cycle time. A search on a larger horizon would reduce the outstanding amount.

5.5.5. *Overview*

In this chapter, we presented a method for structural transformation of a Petri net containing resource shares, inducing structural conflicts. Our goal was to generate a weighted events graph; the deterministic nature of the network then allows an algebraic approach to the study of logical behavior of the obtained Petri net or if unweighted, the application of performance evaluation techniques using dioids, for example (max,+) and (min,+).

The principle of our approach was to take a number of firings from each transition in a cycle into account. These data are used to construct an events graph and establish a system of inequalities, which induces a parameterization through possible initial markings of a weighted events graph. Therefore, the initial Petri net is transformed into a weighted events graph with a set of equations governing the initial marking of newly created places.

The initial data abstraction allows us to consider the potential sequences as a set of solutions of a system of inequalities. The proposed transformation allows us to enrich the linearization techniques already proposed in [JUA 01; TRO 02a] to allow a formal modeling of the behavior of a weighted events graphs with resource sharing using algebra (min, +).

Solving the obtained systems provides a set of possible sequences for a fixed horizon. Evaluating the cycle time of the obtained solutions then provides the opportunity to exhibit an optimal solution with respect to the fixed horizon.

The prospects of this approach are the integration of time constraints in order to obtain qualifying schedules.

5.6. Conclusion

In this chapter, we proposed a mathematical programming model with mixed variables for solving the reachability problem in timed Petri nets. The advantage of our model is that it allows Petri net structures of choice, and is thus able to directly study a non-deterministic model, while guaranteeing the accuracy of the approach. This was validated by numerical experiments using the CPLEX solver with promising results.

In the context of scheduling problems, our approach optimizes the duration of firing sequences leading to the desired marking, and is used to address job-shop problems with blockages, such as the fluidification issues in a railway junction.

The development of cutting and relaxation techniques, dedicated heuristics, as well as identifying relevant lower bounds will also be considered in order to make resolution procedures more efficient.

Our approach can also be extended to problems of formal verification, where the fact of having an incremental search model makes steps explicitly appear and intermediate markings allow us to formulate the properties to check in the form of constraints on the behavior model at the time of construction. For example, by introducing partial markings as in [BOU 04b], between markings whose components are not fully known, we can look for critical scenarios where two places are marked simultaneously, without being concerned with the marking of other places in the Petri net. Thus, we perform a search "guided by the goal" that eliminates some paths of the behavior graph that do not lead to proving the desired property. Therefore, combinatorial explosion issues are handled by the resolution stage itself, for which many efficient tools are being developed.

5.7. Bibliography

[BAC 92] BACCELLI F., COHEN G., OLSDER G.J., QUADRAT J-P., *Synchronization and Linearity - An Algebra for Discrete Event Systems*, Wiley, New York, 1992.

[BEL 82] BELLMANN R., ESOGBUE A.O., NABESHIMA I., *Mathematical Aspects of Scheduling and Applications*, Pergamon Press, 1982.

[BEN 99] BENASSER A., YIM P., "Railway traffic planning with petri nets and constraint programming", *JESA*, 33(8-9), 959–975, 1999.

[BEN 00] BENASSER A., Accessibilité dans les réseaux de Petri: une approche basée sur la programmation par contraintes, PhD Thesis, University of Lille 1, January 2000.

[BER 91] BERTHOMIEU B., DIAZ M., "Modeling and verification of time dependent systems using time petri nets", *IEEE Trans. on Software Eng.*, 17(3), 259–273, 1991.

[BOU 04] BOURDAIS R., TROUILLET B., YIM P., "Étude de nœuds ferroviaires à l'aide des réseaux de Petri temporisés stochastiques", *Conférence Internationale Francophone d'Automatique*, Douz, Tunisia, 2004.

[BOU 04b] BOURDEAUD'HUY T., Techniques d'abstraction pour l'analyse et la synthèse de réseaux de Petri, PhD thesis, École Centrale de Lille, 2004.

[CAM 97] CAMUS H., Conduite de systèmes flexibles de production manufacturière par composition de régimes permanents cycliques: modélisation et évaluation de performances à l'aide des réseaux de Petri, PhD Thesis, University of Lille 1, March 1997.

[CAR 88] CARLIER J., CHRETIENNE P., *Problèmes d'ordonnancement: modélisation / complexité / algorithmes*, Editions Masson, Paris, 1988.

[CHR 82] CHRETIENNE P., CARLIER J., "Les problèmes d'ordonnancement: un domaine très ouvert", *RO-RAIRO*, 175–217, 1982.

[CHR 84a] CHRETIENNE P., "Exécutions contrôlées dans les réseaux de Petri temporisés", *T.S.I.*, 3, 1984.

[CHR 84b] CHRETIENNE P., Les réseaux de Petri temporisés, PhD Thesis, University of Paris VI, 1984.

[COH 89] COHEN G., MOLLER P., QUADRAT J-P., VIOT M., "Algebraic tools for the performance evaluation of discrete event systems", *IEEE Procedings*, 77(1), 39–57, 1989.

[COM 71] COMMONER F., HOLT A.W., EVEN S., PNUELI A., "Marked directed graphs", *Journal of Computer and System Sciences*, 5(5), 511–523, 1971.

[DAM 99] DAMASCENO B.C., Ordonnancement des systèmes de production multi-ressources avec la prise en compte de blocages, PhD Thesis, University of Metz, 1999.

[DAV 92] DAVID R., ALLA H., *Du grafcet aux réseaux de Petri*, Hermes, 1992.

[DEL 03] DELORME X., Modélisation et résolution de problèmes liés à l'exploitation d'infrastructures ferroviaires, PhD Thesis, University of Valenciennes and Hainaut Cambrésis, Valenciennes, 2003.

[DIA 93] DIAZ D., *Les réseaux de Petri - Modèles fondamentaux*, Hermes, 2001.

[GAU 95] GAUBERT S., "Resource optimization and (min,+) spectral theory", *IEEE Transactions on Automatic Control*, 40 (11), 1931–1934, 1995.

[GOT 93] GOThA: Groupe d'Ordonnancement Théorique et Appliqué, "Les problèmes d'ordonnancement", *RO-RAIRO*, 27(1), 77–150, 1993.

[JUA 01] JUANOLE G., VALETTE R., eds., "Sur la modélisation du comportement dynamique des graphes d'événements pondérés", *Modélisation des systèmes réactifs (MSR'2001)*, Hermes, France, October 2001.

[KOR 98] KORBAA O., Commande cyclique des systèmes flexibles de production manufacturière à l'aide des réseaux de Petri: de la planification à l'ordonnancement des régimes transitoires, PhD Thesis, University of Lille 1, 1998.

[LAY 99] LAHAYE S., BOIMOND J.-L., HARDOUIN L., "Timed event graphs with variable resources: asymptotic behavior, representation in (min,+) algebra", *JESA*, 33(8–9): 1015–1032, 1999.

[OHL 94] Ohl H., CAMUS H., CASTELAIN E., GENTINA J.C., "A heuristic algorithm for the computation of cyclic schedules and the necessary WIP to obtain optimal cycle time", *Rensselaer's Fourth International Conference on Computer Intergrated Manafacturing and Automation Technology*, CIMAT'94, vol.1, 1994.

[OHL 95] OHL H., Fonctionnement répétitifs de systèmes flexibles de production manufacturière: analyse et optimisation des performances à l'aide de réseaux de Petri, PhD Thesis, University of Lille 1, 1995.

[RAM 74] RAMCHANDANI C., Analysis of asynchronous concurrent systems by timed Petri nets, PhD Thesis, Massachusetts Institute of Technology, 1974.

[RAU 90] RAUHAMAA M., A comparative study of methods for efficient reachability analysis, Technical Report, Helsinki University of Technology, Digital Systems Laboratory, 1990.

[RIC 97] RICHARD P., Contribution des réseaux de Petri à l'étude de problèmes de recherche opérationnelle, PhD Thesis, University of Tours, 1997.

[SER 89] SERAFINI P., UKOVICH W., "A mathematical model for periodic scheduling problems", *SIAM J. Disc. Math.*, 2 (4), 550–581, 1989.

[SIL 02] SILVA M., GIUA A., COLOM J.-M., eds., "Transformation of cyclic scheduling problem of a large class of FMS into the search of an optimized initial marking of a linearizable weighted T-system", *6th International Workshop on Discrete Event Systems (Wodes'2002)*, IEEE Computer Society, Spain, October 2002.

[TRO 02a] TROUILLET B., BENASSER A., GENTINA J.-C., "On the modeling of the dynamical behavior of weighted t-systems", *APII-JESA*, 36 (7), 931–944, 2002.

[TRO 02b] TROUILLET B., BENASSER A., "Cyclic scheduling problems with assemblies: an approach based to the search of an initial marking in a marked graph", *IEEE international conference on Systems, Man and Cybernetics*, 2002.

List of Authors

Thérèse BONTE
TEMPO Laboratory
Valenciennes University
France

Thomas BOURDEAUD'HUY
LAGIS
Ecole Centrale de Lille
University of Lille Nord de France
France

Alain GIBAUD
TEMPO Laboratory
Valenciennes University
France

Gilles GONCALVES
LGI2A Laboratory
Artois University
University of Lille Nord de France
France

Slim HAMMADI
LAGIS
Ecole Centrale de Lille
University of Lille Nord de France
France

Laetitia JOURDAN
INRIA
Lille Nord-Europe
France

Mostefa Redouane KHOUAHJIA
INRIA
Lille Nord-Europe
France

Mekki KSOURI
ENIT Tunis El Manar University
Tunisia

Frédéric SEMET
LAGIS
Ecole Centrale de Lille
University of Lille Nord de France
France

Christian TAHON
TEMPO Laboratory
Valenciennes University
France

El Ghazali TALBI
INRIA
Lille Nord-Europe
France

Benoît TROUILLET
LAGIS
Ecole Centrale de Lille
University of Lille Nord de France
France

Nawel ZANGAR
Research unit MEDIATRON
SUP'COM
Tunisia

Index

N, O

P, Q, R

S

T, U, V

Lightning Source UK Ltd.
Milton Keynes UK
UKHW04n0733230818
327657UK00008B/55/P

9 781848 214118